MATHEMATICS AS A SCIENCE OF PATTERNS

Mathematics as a Science of Patterns

MICHAEL D. RESNIK

CLARENDON PRESS · OXFORD

OXFORD
UNIVERSITY PRESS

Great Clarendon Street, Oxford OX2 6DP

Oxford University Press is a department of the University of Oxford.
It furthers the University's objective of excellence in research, scholarship,
and education by publishing worldwide in

Oxford New York

Athens Auckland Bangkok Bogotá Buenos Aires Calcutta
Cape Town Chennai Dar es Salaam Delhi Florence Hong Kong Istanbul
Karachi Kuala Lumpur Madrid Melbourne Mexico City Mumbai
Nairobi Paris São Paulo Singapore Taipei Tokyo Toronto Warsaw

with associated companies in Berlin Ibadan

Oxford is a registered trade mark of Oxford University Press
in the UK and in certain other countries

Published in the United States
by Oxford University Press Inc., New York

First published 1997
First issued as paperback 1999

British Library Cataloguing in Publication Data
Data available

Library of Congress Cataloging in Publication Data

Resnik, Michael D.
Mathematics as a science of patterns / by Michael D. Resnik.
Includes bibliographical references.
1. Mathematics—Philosophy. I. Title.
QA8.4.R473 1997 510'.1—dc21 96–51610

ISBN 0–19–823608–5
ISBN 0–19–825014–2 (Pbk)

Printed in Great Britain
on acid-free paper by
Biddles Ltd
Guildford and King's Lynn

I dedicate this book to the memory of my parents
Howard Beck Resnik
Muriel Resnik Jackson

PREFACE

In this book I bring together ideas that I have been developing separately in articles written over the past fifteen years. The book's title expresses my commitment to mathematical realism, empiricism, and structuralism. For in calling mathematics a science I indicate that it has a factual subject-matter and stands epistemically with the other sciences, and in calling it a science of patterns I express my commitment to mathematical structuralism. Contemporary readers in the philosophy of mathematics are likely to know of (if not know) my structuralism and the paper from which the title of this book derives. The same is less likely to hold of my views on realism and the epistemology of mathematics, since much of it appears in conference papers that have not been published, at least not as of this writing. I hope that this book will not only make these newer ideas more readily accessible but also present them and my earlier ideas in a systematic context.

My debt to the writings of W. V. Quine will be apparent to any reader who knows his work. My combination of holism and postulationalism develops the details of Quinean suggestions for an epistemology of mathematics, and his work on ontological relativity has shaped my structuralism.

I am also indebted to a host of individuals for conversations, correspondence and other help. I have acknowledged the help of many of them in previous publications that serve as a basis for this one. I thank them again, but will confine myself to listing only those who have assisted me with this particular manuscript. These are Andrea Bagagiolo, Mark Balaguer, Pieranna Garavaso, Marcus Giaquinto, Eric Heintzberger, Colin McLarty, Geoffrey Sayre-McCord, Adrian Moore, Bijan Parsia, my son David Resnik, Stewart Shapiro, Keith Simmons, and two anonymous referees for Oxford University Press. I am especially grateful to Mark Balaguer and Eric Heintzberger for lengthy commentaries on the previous draft of the book. Angela Blackburn and Peter Momtchiloff in their capacity as philosophy editors of Oxford University Press have encouraged me from the inception of this work, and I thank them both. I also thank Angela

Blackburn for the wonderful job she has done in copy-editing the final manuscript and preparing it for publication.

I am also thankful for two one-semester leaves, one due to a grant from the University of North Carolina Institute for the Arts and Humanities and the other due to the adminstrative grace of Gerald Postema in his capacity as chair of the Philosophy Department.

In writing this book I have drawn from a number of my earlier essays. Some of these have already been published, others are currently in press. In most cases I have substantially rewritten the material in question and interspersed it in various chapters. I thank Oxford University Press and the editor of *Mind* for permission to draw on 'Immanent Truth', which appeared in vol. 99 (1990), and 'A Structuralist's Involvement with Modality', which appeared in vol. 101 (1992). Most of the first paper is reincarnate in Chapter 2 and the Introduction, and sections 1, 3, and 4 of the second paper recur in Chapter 4. I also thank Oxford University Press for 'Ought There To Be One Logic?', which is to appear in Jack Copeland (ed.), *Logic and Reality*, and 'Holistic Mathematics', which is to appear in Matthias Schirn (ed.), *Philosophy of Mathematics Today*. I use material from sections 1, 6, 7, and 8 of the first paper in Chapter 8, and most of the second paper in Chapter 7. I am grateful to the editor of *Noûs* for 'Mathematics as a Science of Patterns: Ontology and Reference', which appeared in vol. 15 (1981), and for 'Mathematics as a Science of Patterns: Epistemology', which appeared in vol. 16 (1982). I use most of the first paper in Chapter 10, and pieces of the second in Chapter 11. I thank the editor and publisher of *Philosophica* for 'A Naturalized Epistemology for a Platonist Mathematical Ontology', which appeared in vol. 43 (1989). I use parts of this paper in Chapters 6 and 9. I thank the editor of *Philosophical Topics* for 'Computation and Mathematical Empiricism', which appeared in vol. 17 (1989), and 'Quine, the Argument from Proxy Functions and Structuralism', which is scheduled to appear in 1997. I use material from the first paper in Chapter 8 and some from the second in Chapter 12. I am grateful to the editor of *Philosophia Mathematica* for 'Scientific vs. Mathematical Realism: The Indispensability Argument', which appeared in 3rd Ser., vol. 3 (1995), and 'Structural Relativity', which appeared in 3rd Ser., vol. 4 (1996). I use most of the first of these articles in Chapter 3, and some of the second in Chapter 12. I thank the Philosophy of Science Association for 'Between Mathematics and

Physics', which appeared in *PSA 1990*, vol. 2. Much of this recurs in Chapter 6. Finally, I thank Routledge Publishing Company for 'Proof as a Source of Proof', which appeared in Michael Detlefsen (ed.), *Proof and Knowledge in Mathematics*. I use parts of this in Chapter 11.

As always, I am indebted to my wife Janet for her encouragement and comfort, and for making life so exciting.

CONTENTS

PART ONE: PROBLEMS AND POSITIONS

PART TWO: NEUTRAL EPISTEMOLOGY

PART THREE: MATHEMATICS AS A SCIENCE OF PATTERNS

PART ONE

Problems and Positions

1

Introduction

Many educated people regard mathematics as our most highly developed science, a paradigm for lesser sciences to emulate. Indeed, the more mathematical a science is the more scientists seem to prize it, and traditionally mathematics has been regarded as the 'Queen of Sciences'. Thus it is ironic that philosophical troubles surface as soon as we inquire about its subject-matter. Mathematics itself says nothing about the metaphysical nature of its objects. It is mute as to whether they are mental or physical, abstract or concrete, causally efficacious or inert. However, mathematics does tell us that its domain is vastly infinite, that there are infinities upon infinities of numbers, sets, functions, spaces, and the like. Thus if we take mathematics at its word, there are too many mathematical objects for it to be plausible that they are all mental or physical. Yet the alternative platonist view that mathematics concerns causally inert objects existing outside space-time seems to preclude any account of how we acquire mathematical knowledge without using some mysterious intellectual intuition.

Resolving this tension between the demands of ontology and epistemology has dominated philosophical thinking about mathematics since Plato's time. Yet after nearly a century of vigorous work in the foundations and philosophy of mathematics the problem remains as acute as ever. For we have a greater appreciation than previous generations of philosophers of the boundlessness of the mathematical universe and the mathematical requirements of science. Rigorous reflections on the great, but unsuccessful, attempts by Frege, Hilbert, and Brouwer to work out philosophically motivated foundations for mathematics have shown us exactly why it will not do to take mathematics to be an a priori science of mental constructions, or an empirical investigation of the properties of ordinary physical objects, or a highly developed branch of logic or a game of symbol manipulation. On the other hand, the naturalism driving contemporary epistemology and cognitive psychology demands that we not settle for an account of mathematical knowledge based upon processes, such as a priori intuition, that do not seem to be capable of scientific investigation or explanation.

This has led many contemporary philosophers of mathematics to disdain realism about mathematical objects, and to not read mathematics at face value. Hartry Field has embraced an ingenious version of the view that mathematics is a useful fiction. Geoffrey Hellman exchanges realism about mathematical objects for realism about possible ways unspecified objects might be related to each other. Charles Chihara reads mathematical existence statements as asserting that certain inscriptions are possible. Efforts towards fully formulating these views have produced impressive formalisms. But, to make a point I will argue later, the epistemic and ontic gains these approaches promise prove illusory when applied to the infinities found in the higher reaches of contemporary mathematics.

I mention these anti-realists now by way of background to my main project in this book, which is to defend a version of mathematical realism. In the next paragraphs I sketch the view I will amplify in subsequent chapters.

My realism consists in three theses: (1) that mathematical objects exist independently of us and our constructions, (2) that much of contemporary mathematics is true, and (3) that mathematical truths obtain independently of our beliefs, theories, and proofs. I have used the qualifier 'much' in (2), because I do not think mathematical realists need be committed to every assertion of contemporary mathematics. At a minimum, realists seem to be committed to classical number theory and analysis, for less than this opens the way to anti-realist, constructive accounts of mathematics. Moreover, accepting classical analysis already suffices for making a convincing case that the mathematical realm is independent of us and our mental life, thereby raising epistemological problems for realists. I am inclined to commit myself to standard set theory as well, but the evidence for this much mathematics and beyond is not as firm as it is for analysis and number theory, and, as a result, the case for a realist stance toward it is weaker.

The ontological component of my realism is a form of structuralism. Mathematical objects are featureless, abstract positions in structures (or more suggestively, patterns); my paradigm mathematical objects are geometric points, whose identities are fixed only through their relationships to each other. This structuralism explains some puzzling features of mathematics: why it only characterizes its objects 'up to isomorphism', and why it may use alternative definitions of, say, the real numbers, when these definitions are not even extensionally equivalent. Yet structuralism also yields a form of ontological relativity: in certain contexts there is no fact as to whether, say, the real numbers are the points on a given line in Euclidean space or Dedekind cuts of rationals. And this will require some explaining.

Material bodies in various arrangements 'fit' simple patterns, and in so doing they 'fill' the positions of simple mathematical structures. We may well perceive such arrangements, but we do not perceive the positions, the mathematical objects, themselves; since, on my view, they are not spatiotemporal. How then did we come to form beliefs about them—short of using the sort of non-natural or a priori processes I renounce? I hypothesize that using concretely written diagrams to represent and design patterned objects, such as temples, bounded fields, and carts, eventually led our mathematical ancestors to posit geometric objects as *sui generis*. With this giant step behind them it was and has been relatively easy for subsequent mathematicians to enlarge and enrich the structures they knew, and to postulate entirely new ones.

Basing the epistemology of mathematical objects on positing has the advantage of appealing to an apparently natural process akin to making up a story. However, it also generates the obvious problem of showing how positing mathematical objects can lead us to mathematical truths and knowledge. Clearly mere originality would not be enough to justify our ancestors' initially suggesting that mathematical objects exist, much less retaining them in their conceptual scheme. I believe they were justified in introducing mathematical objects because doing so promised to solve a number of problems confronting them and to open many new avenues of thought. Part of their (and our) justification for retaining mathematical objects was (and remains) pragmatic and global: they have proved immensely fruitful for science, technology, and practical life, and doing without them is now (virtually) impossible.

Scientists also posit new entities, ranging from planets to subatomic particles, and they have done this with great effect. For the most part, however, they posit to explain observable features of the world in causal terms, and they usually insist upon experimentally detecting their posits. This is very unlike mathematics. Furthermore, when scientists relax the detection requirement, they sometimes own that the entities in question are merely fictional idealizations. This prompts the worry that science also regards mathematical objects as merely the fictitious characters of a useful and powerful idealization. However, as I shall show, a careful analysis of the way scientists use mathematics reveals that they presuppose its truth. Even when using such devices as point-masses, frictionless objects, or ideal gases to develop idealized models, they presuppose the reality of the mathematical objects to which they refer. One philosophical consequence of this is that certain anti-realists in the philosophy of science are still committed to the reality of mathematical objects.

Although my argument from the role of mathematics in science forestalls the fictionalist's ploy, it generates a new worry: namely, that we may not be justified in accepting a mathematical claim unless it is presupposed by science. If this is true, we should suspend judgement on much contemporary set theory—an unsettling consequence for many mathematical realists.

In practice, when justifying a mathematical claim we hardly ever invoke such global considerations as the benefits to natural science. We ordinarily argue for pieces of mathematics locally by appealing to purely mathematical considerations. Proving theorems is an obvious way, but one that passes the buck to the axioms. Usually we accept axioms because they yield an important body of theorems or are universally acknowledged and used by practising mathematicians. These are considerations restricted to mathematics and its practice proper. They form part of a local conception of mathematical evidence; and we can invoke this to support some of the mathematics that currently has no use in natural science.

It would be wrong to conclude from its possessing a local conception of evidence that mathematics is an a priori science, disconnected evidentially from both natural science and observation. First, observation is relevant to mathematics, because when supplemented with appropriate auxiliary hypotheses, mathematical claims yield results about concretely instantiated structures, such as computers, paper and pencil computations, or drawn geometric figures, that can be tested observationally in the same way that we test other scientific claims. Secondly, technological and scientific success forms a vital part of our justification for believing the more interesting parts of mathematics, the parts that go beyond the computationally verifiable.

I formulate my mathematical realism and its epistemology using notions of truth and reference that are immanent and disquotational. This means that they apply only to our own language, and serve primarily to permit inferences such as the following: 'Everything Tess said is true, and she said "Jones was at home"; so Jones was at home.' Even this modest conception of truth allows me to formulate theses committing me to an independent mathematical reality. Moreover, it avoids worries about how our mathematical terms 'hook onto' mathematical objects, and permits me to explain how merely positing mathematical objects, objects to which we have no causal connection, can enable us to refer to and describe them.

Structuralism also enters my epistemology at a number of points. As I already mentioned, it is part of my account of the genesis of mathematical knowledge. It also figures in my explanation of how manipulations with concrete numerals and diagrams can shed light on the abstract realm of mathematical objects. Here the key idea consists in noting that these concrete devices represent the abstract structures under study. The unary numerals, for example, and the computational devices built upon them, reflect the structure of the positive integers.

Which structures we recognize will depend upon how finely grained a conception of structure we use. This in turn is a function of the devices we recognize for delineating structure; and, according to many contemporary approaches to structure, this ultimately turns on where we set the limits of logic and logical form.

Once one identifies structure with logical form it is a short step to thinking that there must be just one correct conception of structure. My views on both logic and structure contravene this. First, I hold that in calling a truth a logical truth we are not ascribing a property to it that is independent of our inferential practices, such as *being true in virtue of its form*. What we count as logically true is a matter of convenience. Consequently so are the limits of logic and logical form. Secondly, structural similarity is like any another similarity; it presupposes a respect in which things are to be compared. Two things can be structurally similar in one respect and not in another, for example, the same in shape but not in size. Thus structure is relative to our devices for depicting it. But this does not undercut my realism, since the facts about structures of a given type obtain independently of our recognizing or proving them.

Here is the plan of the book. Part One: 'Problems and Positions' begins by explaining my mathematical realism and the version of truth I use. Next is a chapter offering a prima facie case for mathematical realism, which is based largely on an argument that the indispensability of mathematics to science justifies a realist stance towards much mathematics. This is followed by a critique of anti-realist work by Charles Chihara, Hartry Field, Geoffrey Hellman, and Philip Kitcher aimed at undermining the indispensability premiss upon which the argument is based, and then by a review of the epistemic problems motivating anti-realists.

In Part Two: 'Neutral Epistemology' I take up issues in the epistemology of mathematics that I can treat independently of my structuralist ontological doctrines. I begin with a critique of the distinction between mathematical and physical objects, since this underlies almost all contemporary thinking about the epistemology of mathematics. I turn then to a holist approach to the epistemology of science and explain how this can be compatible with defeasible local conceptions of evidence operating in the various branches of science. This makes room for a local conception of mathematical evidence. However, knowledge based upon such evidence fails to be a priori, because, in principle, observational evidence bearing upon scientific systems containing mathematical claims can provide a basis for overriding the mathematical evidence for those claims. Further analysis shows that even our local conception of mathematical evidence recognizes the relevance of empirical data. Since deduction plays such an important role in the methodology of mathematics, I turn next to the nature of logic, where I argue against realism concerning logical necessity and possibility. My position again has an anti-apriorist slant, since I hold that the role of logic in mathematics is purely normative—guiding inference rather than reporting so-called logical facts.

I then move from questions of justification to questions about the genesis of mathematical knowledge. Here I begin to develop the view that mathematical objects are posits. Taking mathematical objects as posits raises the question of how and in what sense our beliefs can be about mathematical objects. I argue that the sense in question can be handled by an immanent, disquotational approach to reference.

To complete my epistemology of mathematics I bring in my structuralist account of mathematics, the focus of Part Three:

'Mathematics as a Science of Patterns'. Here I expound a theory of patterns (structures) and argue that we can resolve a number of issues in the ontology of mathematics by construing mathematical objects as positions in patterns. I also argue that mathematical knowledge has its roots in pattern recognition and representation, and that manipulating representations of patterns provides the connection between the mathematical proof and mathematical truth. I conclude this part by addressing issues concerning structuralism itself—including the relationship between logical form and structure, and the possibility of a structuralist foundation for mathematics.

2

What is Mathematical Realism?

The view I will propose and defend is a form of mathematical realism. Now just calling my view *realism* threatens to subject it to gratuitous objections—which I could easily avoid by not labelling it at all. To make matters worse, few philosophical terms are currently more controversial or obscure. But labels help locate views on the philosophical map and indicate whether some clash more severely than others. My view is opposed to views about mathematics that claim to be anti-realist and go by such names as 'nominalism', 'constructivism', 'fictionalism', 'deflationism'; in this respect 'realism' fits. My view also has much in common with other so-called realist views in other areas of philosophy. So I am going to retain the label, and try to define it so that it characterizes the contemporary debate about mathematical objects as well as traditional debates between realists and anti-realists in other areas of philosophy.[1]

1. TO CHARACTERIZE REALISM . . .

One may be a realist about some things without being a realist about others. For instance, one might believe in the existence of physical objects while denying the reality of mental or abstract objects. Thus realism is not a single view but rather a family or collection of views. One is not a realist *simpliciter* but rather a realist with regard to *X*s, where *X*s might be mathematical objects, electrons, propositions, moral values, and so on.

Despite the variety of realisms, surveying traditional debates between realists and anti-realists reveals that three themes are likely to emerge as part of a realist's position: an existence theme, a truth

[1] Call me a 'platonist', if you like. I used this label in my earlier writings. But I am using the term 'realism' since many of the contemporary philosophers with whom I debate or ally myself use it. See e.g. Maddy (1990) and Field (1988).

theme, and an independence theme. Realists accept certain entities (for example, material objects, electrons, universals, sets, possible worlds, beliefs); while anti-realists reject them outright, or else, in order to retain the *appearance* of reference to the entities in dispute, they substitute other entities for them (for example, 'logical constructions' of sense data for material objects, predicates for properties, maximally consistent sets of sentences for possible worlds). Realists also regard the entities they accept as having properties, as standing in relations, and as giving rise to facts. Many realists add that our current theory about these entities is approximately, basically, largely, or unqualifiedly true.[2]

But there is more. Berkeley denied neither the existence of tables and chairs nor most of our beliefs about them, but he did deny that they exist independently of the mind of God and he regarded them as collections of ideas. Some mathematical anti-realists, while denying the outright existence of mathematical infinities, retain them in the form of possibilities involving *concreta*. Thus issues concerning independence play as important a role in debates between realists and anti-realists as issues concerning existence and truth.

Let us then try this schematic formulation of Realism. One is an realist with regard to *X*s just in case one holds these three theses about *X*s: (1) *X*s exist, (2) our current theory of *X*s is true (approximately, basically or largely true), and (3) the existence of the *X*s and the truth of statements about *X*s is somehow independent of us (or perhaps other entities, depending upon the realist/anti-realist debate in question).

In striving for generality, I have used vague, hedging terms in conditions (2) and (3). The exact way in which they might be more fully specified varies with the branch of philosophy in which the realism/anti-realism debate occurs. Consequently, I will only attempt to do this (below) for the case of philosophy of mathematics.

It also turns out that conditions (1), (2), and (3) are not *jointly* necessary for each type of realism; nor is any one taken *alone* sufficient for each type. For example, it does not seem essential to realism about value that abstract values, such as Goodness, exist, although it

[2] In recent years the truth of contemporary science has become an important component of the debates concerning scientific realism. Nancy Cartwright and Ian Hacking, so-called entity realists, countenance subatomic particles while denying the fundamental scientific theories describing them. See Cartwright (1983) and Hacking (1982).

does seem essential that value judgements be true or false independently of our wishes, conventions, practices, and so on.

Thus (1) is not necessary for all forms of realism. It is not sufficient either. We have already noted that Berkeley did not deny the existence of tables and chairs. And asserting that numbers exist as mental constructions certainly does not make one a mathematical realist.

The same situation arises concerning condition (2). Nobody denies the reality of lightning. Yet scientists are unwilling to affirm any of the current theories of its nature, since each fails to answer important questions about how lightning is generated. Holding that our current theory of *X*s is true (or qualifiedly true) need not suffice for realism about *X*s either. Certain nominalists grant that number theory is true, but then they disavow numbers by adding that, properly understood, number theory concerns possibilities for inscribing numerals.

Consider next condition (3), that the existence of *X*s and the truth or falsity of statements about *X*s must be independent of us (or other appropriate entities). Elliot Sober has remarked that this condition encounters some obvious but serious counterexamples.[3] Surely we cannot require realists about human beliefs to hold that our beliefs exist independently of us, or that the truth or falsity of our theory of what we happen to believe is independent of what we happen to believe. On the other hand, condition (3) alone also fails to suffice for certain forms of realism. One might hold that the existence of minds, for example, or numbers, or electrons, as well as the truth of our theory of them, is an entirely objective matter, independent of what we happen to believe and of the evidence we happen to possess, and then go on to deny minds, numbers, or electrons and theories affirming their existence. Hartry Field, a prominent mathematical anti-realist, takes exactly such a position concerning numbers, as does Bas van Fraassen, an equally prominent scientific anti-realist, concerning unobservables.[4]

I hypothesize that if we examined the realism/anti-realism debates in various branches of philosophy, we would find that one cannot be a realist about any kind of thing unless one maintains at least one of (1)–(3) with respect to those things. Furthermore, we

[3] See Sober (1982).

[4] Field (1988), van Fraassen (1980).

would surely count as a realist concerning some things anyone who maintained all three conditions with respect to those things. So we seem to have found a necessary condition for realism, and a sufficient one, but none which is both necessary and sufficient. Classifying a philosophical view as realist is like diagnosing an illness as a case of arthritis, lupus, or schizophrenia. For these and a number of other diseases there are no fixed diagnostic criteria. Instead the appropriate medical associations, such as the American Rheumatism Association, have established lists of criteria for use in diagnosing patients. If the patient meets more than a certain pre-set number on the list, the presumption is that the diagnosis is positive; otherwise there is no such presumption. The criteria are to be used with caution and tempered by the circumstances of the case. In a similar way, the three conditions we have been examining are useful criteria for classifying metaphysical positions. However, in using them one should always proceed on a case-by-case basis, paying particular attention to the philosophical context or debate in which the position is found.[5]

[5] The characterization of realism offered here has a well-known rival in Michael Dummett's view that realists concerning a given discourse maintain (in opposition to anti-realists) that to understand a sentence in the discourse is to know what conditions obtain if it is true, and that such a sentence has a truth-value independently of our ability to verify it. (I have taken this characterization from Dummett (1975).) One reason that I do not use his criterion is that it counts as realists many who call themselves as anti-realists. Examples are Hartry Field (in Field (1988)) who holds that contemporary mathematics is false, and Charles Chihara (in Chihara (1990) who holds that it is true when properly interpreted in his constructibility theory. Bas van Fraassen (van Fraassen (1980)) is a well-known, non-verificationist, anti-realist in the philosophy of science. I discuss Field's and Chihara's views more fully in subsequent chapters.

Another rival criterion is due to Geoffrey Sayre-McCord (Sayre-McCord (1988)) who holds that in any debate between realists and anti-realists there will be a disputed class of sentences; realists will hold that some members of this class are literally true when literally construed; anti-realists will deny this. Despite its economy and elegance, I demur at Sayre-McCord's criterion. It would count Brouwer as a realist, although even he regarded himself as an idealist. For he claimed that some existential mathematical sentences are literally true, maintaining that on a literal construal they report the results of private mathematical constructions carried out in inner intuition. Of course, when compared to those formalists who take mathematics to be a contentless game, Brouwer comes out on the realist side. Classifying Brouwer as more realist than game formalists accords with both Sayre-McCord's criterion and the opinions of some philosophers of mathematics. It may be, then, that Sayre-McCord's criterion will be useful in marking out a spectrum of positions ranging from radical to moderate anti-realism and thence to realism proper. In any case, I will stick with my more stringent condition (2) and less elegant set of criteria.

In characterizing mathematical realism we require all three of conditions (1)–(3). Intuitionists hold that (certain) mathematical objects exist and thus meet condition (1). But they meet neither (2) nor (3). For they hold that mathematical existence and truth depend upon our constructions and proofs, and they also reject large portions of contemporary mathematics. Hartry Field's view, as we have noted, meets condition (3) but neither (1) nor (2). Finally, the recent modal interpretations of mathematics, such as Charles Chihara's and Geoffrey Hellman's, claim to account for the truth and independence of contemporary mathematics without having to acknowledge the existence of mathematical objects. These views meet (2) and (3) but not (1). Thus it is necessary to affirm all three conditions to be a mathematical realist. To the best of my knowledge they are jointly sufficient; I cannot think of any acknowledged mathematical anti-realism that satisfies all three.

2. IMMANENT TRUTH

Talk of truth plays a major role in formulating realism. I have reflected this in conditions (2) and (3), but I have not explained the conception of truth presupposed in those conditions. This is not something to be set aside, since some important recent general critiques of realism have concentrated upon the correspondence conception of truth that realism is presumed to presuppose. These critiques are misguided, because a weaker, but none the less non-epistemic, conception of truth suffices for realism. The purpose of this section is to explain this conception of truth and to show that it suffices for mathematical realism.

Now, in searching for an adequate conception of truth, a philosopher can hardly fail to fall under the spell of one our most serious epistemic predicaments. On the one hand, only from within the perspective of our conceptual scheme can we judge something to be true. On the other hand, we know that we will significantly revise this very scheme, and thereby reject much of that we now rationally accept as true. The first half of this predicament pulls philosophers towards an epistemic conception; the second towards a correspondence conception. Yet a satisfactory resolution lies on neither extreme. The difficulties with characterizing truth via a correspondence with reality are well-known: It is not clear in what the correspondence consists—that is, which parts of language or thought correspond to which parts of reality. Nor is it clear whether the correspondence must be unique, or how it could be established at the outset, since any correspondence we set up using a bit of language would seem to presuppose a prior correspondence for that bit of language. Yet the usual epistemic conceptions face an equally serious obstacle. Because they are stated in terms of epistemic idealizations, such as the final scientific theory or ideal warranted assertability, they do not apply to our own epistemically flawed languages and theories unless they are supplemented with controversial premisses. For example, because we don't know whether 'People act on their desires' is even a sentence of the final scientific theory, we have no grounds for inferring the biconditional:

> 'People act on their desires' is true (in the sense of being asserted by the final scientific theory) if and only if people act on their desires.

Yet, as we will see below, such biconditionals fund some of our most important inferences involving truth.

The conception of truth I will expound, one I will call an *immanent conception of truth*, avoids the problems of epistemic and correspondence approaches while applying directly to sentences in our own language. When coupled with classical logic, it also furnishes the familiar principles linking the truth-values of compound sentences to those of their components, and yields the law of bivalence.[6] Furthermore, it makes room for our most fundamental realist intuitions by permitting truth to be independent of our present theories and methods. Despite this, it does not generate worries about how a sentence corresponds to reality or how it is related to what might be affirmed under ideal epistemic conditions.

2.1. *Truth Vehicles, Truth-Theories, and Some Conceptions of Truth*

People philosophizing about truth encounter a stumbling-block at the outset, since they must choose their truth vehicles, that is, the things they take to be true or false. One gets a cleaner theory if nothing is both true and false, but the price is not cheap. Neither sen-

[6] Although I here take implying bivalence in the presence of classical logic to be an advantage of this theory, in Chapter 12 I will propose restricting bivalence and classical logic to resolve problems they raise in connection with my structuralism.

tences *qua* linguistic patterns nor *qua* specific utterances can be guaranteed to work, due to ambiguities that can even survive the contexts of specific utterances and speakers' intentions. Although ambiguity is prevalent in non-scientific discourse, even mathematics is not immune to the problem: '1 + 1 = 1', for example, is true when construed as a statement of Boolean algebra.

Thus it has been common for philosophers to postulate propositions, understood as sentential meanings, to serve as truth vehicles. The idea here is that an ambiguous utterance may express several propositions, but none of these can be both true and false. But despite the many reasons one might cite for appealing to propositions in philosophizing about mind and language, they are probably even more controversial than mathematical entities. This makes them a poor beginning for a defence of mathematical realism.[7]

Now the only reason that mathematical realists need worry about truth is that they want to affirm that various mathematical theories are true. Although these theories are often formulated in natural languages or their technical extensions, their sentences are, for the most part, not irredeemably ambiguous or context-dependent; with notational changes they could be expressed so that they had at most one truth-value. I do not think it is an unreasonable idealization to assume that this has been done, and I will do so. We can restrict ourselves to a collection of mathematical languages containing only unambiguous declarative sentences none of which occurs in more than one of these languages. Then we can take the sentences of these languages as our truth vehicles.[8]

[7] And, perhaps, a question-begging one. For since propositions are abstract entities, given enough of them, it is likely that we could reduce mathematical objects to propositions.

[8] So long as we take for granted that each sentence has at most one truth-value, many of the remarks to follow apply to truth-theories for languages that are less restricted than the ideal mathematical languages considered here.

I also realize that I am 'assuming away' some of the major difficulties with taking sentences as truth vehicles. However, taking propositions as truth vehicles only seems to transfer these difficulties from a truth-theory to its applications. Rather than restricting our truth-theories to languages in which no sentence is both true and false we must restrict their applications to languages in which no sentence expresses both a true and a false proposition (or fails to correspond to an unique part of reality). Either way we must face the preparatory work of identifying declarative (or truth-apt) sentences, disambiguating them and removing their context dependencies, indexicals, and the like. At least in the mathematical case this task is less daunting than it is for language at large. By the way, I am not claiming that these considerations show that we have no need for propositions in other areas of philosophy.

I will also identify theories with collections of sentences, and count a theory true when and only when all its sentences are true. By a *truth-theory* for a given language or languages, I shall mean a theory containing at least some assertions to the effect that specific sentences of the language or languages are true under certain conditions. Truth-theories may also contain generalizations about the conditions under which sentences are true, but they need not. They may apply to all sentences of the language(s) in question, but they might be restricted to portions of the language(s). Truth-theories may also differ radically in the devices they use for specifying the conditions under which a sentence is true.

Since I will largely ignore the details of the various truth-theories that I will consider, I will speak of *conceptions* of truth to indicate families of related truth-theories. I like to think of a particular conception of truth, for example, the correspondence conception, as a *way of thinking* about truth—something guiding the construction of particular truth-theories or exhibited in them. Thus a variety of truth-theories may develop a particular conception of truth. For example, one correspondence truth-theory may only admit correspondences based upon causal chains, another may admit those based upon intellectual intuitions; one truth-theory may refer to the correspondence relation explicitly through its singular terms, another may only use a correspondence predicate. Similarly, one epistemic truth-theory might count a sentence as true if it is rationally acceptable by current standards, another only if it has been conclusively verified, a third if and only if the final science affirms it.

Construing conceptions of truth as ways of thinking about truth is useful for highlighting the considerations motivating different truth-theories, but it tends to produce characterizations too vague for assessing the commitments of a given conception of truth. So I will try to give more precise definitions of the various conceptions of truth by defining them in terms of conditions that truth-theories answering them must meet.

I will count a truth-theory as meeting the *correspondence conception* just in case (a) it specifies a word–world reference relation (a relation with an argument place for expressions and one for objects generally), and (b) for each sentence *S* to which the truth-theory applies, it implies a sentence of the form

S is true if and only if ——,

where the blank is replaced by a condition on *S* formulated in terms of its sentential structure and the reference relation specified by the theory.[9]

Unfortunately, this definition is not satisfactory as it now stands. According to this definition, truth-theories based upon *disquotationally defined* reference relations (for example, the relation defined as what *t* bears to *x* when *t* is 'Adam' and *x* is Adam or *t* is 'Beatrice' and *x* is Beatrice, and so on) count as correspondence theories right along with those based upon, say, causally defined reference relations. Disquotationally defined reference relations are word–world relations: they relate words to objects, and thereby satisfy clause (b). Yet truth-theories based upon them hardly conform to the correspondence way of thinking about truth. Correspondence theorists seek a completely general account of truth, one attributing truth-conditions even to sentences which we do not understand and cannot translate, one purporting to explain what truth is. By contrast, truth-theories based upon disquotationally defined reference simply specify the extension of the term 'true' for the language to which they apply. Furthermore, to apply them one must already know the referents of the terms in the object language, for the definitions of disquotationally defined reference relations *use* these terms to specify their own references. The truth-theories based upon these relations provide no general account of truth or reference.

For similar reasons, theories that characterize truth in terms of translation plus a disquotationally defined reference (for example, by first defining 'true' disquotationally for their home languages and then extending its application to foreign sentences by stipulating that a foreign sentence counts as true just in case its home translation is true) are inimical to the correspondence way of thinking about truth. These theories can furnish considerably more general truth and reference predicates than purely disquotational theories, but by virtue of their ultimate appeal to disquotationally defined reference relations they also fail to explain what truth is for arbitrary languages, which is one the main goals of correspondence theorists.

In view of this we must add another clause to (a) and (b) above in order to exclude disquotationally based truth-theories. The following condition suffices:

[9] This formulation is not as general as Marian David's, but it will serve our purposes here. See David (1994).

(c) the word–world reference relation through which the truth-theory in question characterizes truth must apply to words in arbitrary languages.

Adding this clause excludes disquotationally based theories, because disquotationally defined reference relates only a fixed set of terms to their referents. On the other hand, (c) does not exclude truth-theories based upon reference relations specified through conditions mentioning causal chains, intentions, platonic intuitions, or physically described behaviour, since such conditions (presumably) apply to arbitrary languages.

I will count a truth-theory as answering to the *epistemic conception* just in case for each sentence *S* to which the truth-theory applies, it implies a sentence of the form

S is true if and only if ——

where the blank is replaced by an epistemic condition on *S*, such as: being assertible, justified, verified, or warranted, belonging to an epistemically ideal theory, or warranted under epistemically ideal conditions. This characterization is vague—at least to the extent that the term 'epistemic condition' is. Yet it is clear enough for us to see that the epistemic and correspondence conceptions need not determine the same class of truth-theories. Epistemic truth-theories need not imply biconditionals depicting truth in terms of a reference relation; correspondence theories need not imply epistemic biconditionals applying to whole sentences.

Despite this, one and the same truth-theory might satisfy both conceptions. For example, a truth-theory for a language containing just observation sentences might define truth using a causally specified reference relation, and contain a theorem stating that such sentences are true if and only if they are verifiable. At a less fanciful level, it is easy to develop a truth-theory for effectively decidable number-theoretic sentences using a set-theoretically specified reference relation, and prove within the theory that truth for such sentences is coextensive with their provability within some standard system of number theory.

2.2. *Disquotational Biconditionals for Truth*

Since Alfred Tarski's pioneering work on defining truth nearly everyone who writes about truth holds that a proper truth-theory must imply a disquotational biconditional for truth for each sentence within its scope. (In speaking of the theory having a scope, I am allowing that it may apply to only part of a language or to more than one language.) By a *disquotational biconditional* for truth I mean a sentence of the form

'*p*' is true (in L) if and only *p*.[10]

People usually do not explain why truth-theories should meet this requirement, so I am going to devote the next few paragraphs to presenting what I take to be the strongest reason in its favour.[11]

Consider the following patterns of reasoning:

Pattern (1): *p*; because according to (a true) theory *K*, *p*;
Pattern (2): Theory *K* is not true; because *K* is true only if *p*, and not *p*.

The pattern (1) typifies inferences in which we infer a particular statement by asserting a collection of statements that imply it, while (2) typifies those in which we reject a collection of statements on the grounds that they imply a particular statement that we reject. It is important to note that whether we be realists or anti-realists, whether we like correspondence truth, epistemic truth, or neither, we will want to validate inferences of these types.

This is easy enough when the theory *K* in question can be codified by means of a finite number of axioms. For then to affirm or deny *K* one need only affirm or deny the conjunction of its axioms, since this implies each of *K*'s assertions and is in turn implied by the collection of them. This will allow us to replace patterns (1) and (2) above by:

Pattern (1a): *p*; because *q* & *r* & . . . & *s*, and if so, *p*;
Pattern (2a): not (*q* & *r* & . . . & *s*); because *q* & *r* & . . . & *s* only if *p*, but not *p*,

[10] I use the letters '*p*' and '*q*' as *schematic* letters standing in place of sentences and '*S*' as a *variable* ranging over sentences. Thus the displayed schema represents sentences such as:

'10 + 3 = 13' is true if and only if 10 + 3 = 13,

but not the sentence:

'*p*' is true if and only if 10 + 3 = 13.

[11] A number of the logical points I make about truth, especially those in this section, draw upon Field (1986).

and avoid truth talk altogether. One can use this method for asserting or denying theories, so long as they are finitely axiomatized.

Of course, this technique fails for theories which are not finitely axiomatizable. But suppose that we spoke a language with a device for forming infinite conjunctions. Then we could replace (1) and (2) by:

> Pattern (1b): *p*; because InfConj *C*, and if so, *p*;
> Pattern (2b): not InfConj *C*; because InfConj *C* only if *p*, but not *p*.

In such languages we could also affirm or deny theories and carry out inferences conforming to (1) and (2) by asserting or denying the appropriate infinite conjunctions.

In languages such as ours we use a truth predicate instead. We assert a theory *K* by asserting that all its sentences are true. We deny it by denying at least one of its sentences. This lets us spell out (1) and (2) as:

> Pattern (1c): *p*; because '*p*' belongs to *K* and all *K*'s sentences are true;
> Pattern (2c): *K* is false; because '*p*' belongs to *K*, and not *p*.

Notice that these will not do the work of (1) and (2) unless we assume disquotational biconditionals for each '*p*' to which we apply the above schemata. For to pass from '"*p*" is true' ('"*p*" is false') to '*p*'('not *p*') and back, we require the biconditional:

> '*p*' is true if and only if *p*.

In short, *no truth predicate can replace infinite conjunction unless we can use it to disquote the sentences to which it applies*.

Given the importance of the disquotational biconditionals to inferences involving the predicate 'true', one would expect that every conception of truth would hold them to be essential to its truth-theories. Indeed, many philosophers take such biconditionals to be *definitive* of correspondence theories. But this is wrong. Disquotational biconditionals alone do not make a theory into a correspondence theory; other theories—even epistemic theories—can imply them too. Take, for example, a metatheory ML for a language L having a syntactically complete and consistent proof procedure. Add to ML the following rule of inference: From '"*p*" is provable in L' infer *p*. For each sentence of L we can now prove in ML an instance of the schema:

> *p* if and only if '*p*' is provable in L.[12]

Thus if in ML we define 'true (in L)' as 'provable in L', we can prove all the disquotational biconditionals required for the ML account of truth for L. Yet plainly, as it stands, this ML does not answer to the correspondence conception, since it characterizes truth in purely syntactic terms.

What is more, a correspondence theory might not imply the requisite disquotational biconditionals on its own. Of course, both correspondence and epistemic theories imply biconditionals of the form:

> (a) '*p*' is true if and only if C('*p*'),

where 'C(*x*)' is a predicate applying to sentences, such as 'is warranted' or 'is composed of terms referring to so and so'. Yet *taken by themselves* such sentences fail to imply ones of the form:

> (b) C('*p*') if and only if *p*,

and consequently those of the form:

> (c) '*p*' is true if and only if *p*.

Depending on one's approach to truth, the extra premisses needed to fill this gap can take quite different forms, ranging from the trivial to the substantial.[13]

Many contemporary correspondence theorists expect to use a causal account of reference to fill the gap. They aim to follow Tarski to the extent of characterizing truth recursively in terms of word-reference and then to depart from him by explicating word-reference

[12] Proof: Suppose *p*. Then if '*p*' is not provable in L, 'not *p*' is by the syntactic completeness of the proof procedure. But then by the inference rule, not *p*. From this contradiction we infer that '*p*' is provable in L. The converse is an immediate application of the inference rule.

[13] Suppose that a correspondence theory implies just biconditionals of the form:

> *S* is true iff and only if *S* corresponds to the proposition that *p* and *p*.

Then to go from, say, 'snow is white' to '"snow is white" is true', we would need additional premisses to show that the sentence 'snow is white' corresponds to the proposition that snow is white.

causally.[14] Assuming that the latter part of this programme succeeds, the resulting truth-theory will make assertions of the form:

> Expression *e* refers to object *o* if and only if *e* and *o* stand in causal relation *R*.

But these assertions alone do not suffice for the disquotational biconditionals for truth; we need also corresponding disquotational biconditionals *for reference*. For example, it is not enough to know that

> 'George Washington' refers to the object to which it bears relation *R*,

to conclude, via Tarski's recursions, that

> 'George Washington lived in America' is true if and only if George Washington lived in America.

We must also know that the object to which the name 'George Washington' bears *R* is George Washington.

(It has never been entirely clear to me how causal theories will furnish information of this kind, but I presume that it is supposed to follow unproblematically from an account of our naming practices.)

Obtaining disquotational biconditionals is no easy task for those epistemic truth-theorists who apply their truth-predicates to undecided sentences. Suppose, for example, one defines (epistemic) truth as idealized rational acceptability. How does one show that if '*p*' would be accepted by an ideally rational agent, then *p*? (Here both the occurrences of '*p*' are to be replaced by the same sentence.) In particular, how does one prove this generally without employing a non-epistemic notion of truth?

This not to say that one can never plausibly derive disquotational biconditionals from an epistemic theory of truth. Earlier we saw that this can be plausibly done for a language having a syntactically complete and consistent proof procedure. Finally, the questions I raised in the last paragraph fall short of refuting epistemic approaches to truth, as do my prior observations on contemporary correspondence theories. I have only aimed here to point out difficulties which both approaches to truth must address.

One way to ensure that a truth-theory yields enough disquota-

[14] See, for example, Devitt (1984).

tional biconditionals is to formulate it in a metalanguage containing a device for forming infinite conjunctions, as well as a postulate or theorem stating an equivalence between one's truth-predicate and an infinite conjunction implying the requisite disquotational biconditionals for truth. Let me illustrate the idea with a hypothetical language containing just three sentences, A, B, and C. The truth-theory for this language should yield three disquotational biconditionals:

'A' is true if and only if A;
'B' is true if and only if B;
'C' is true if and only if C.

Now suppose that in constructing this theory we stipulate (either as an axiom or as a definition) the following:

(T) *x* is true if and only if
x = 'A' only if A, and
x = 'B' only if B, and
x = 'C' only if C.

Then we can derive each of the disquotational biconditionals above from (T) and some obvious principles of syntax and logic. Here is the derivation for A, using two conditional proofs, (a)–(c) and (d)–(h), to obtain the biconditional (i):

(a) 'A' is true (assumption)
(b) 'A' = 'A' only if A
(c) A

(d) A (assumption)
(e) 'A' = 'A' only if A
(f) 'A' = 'B' only if B
(g) 'A' = 'C' only if C
(h) 'A' is true

(i) 'A' is true if and only if A.

(Step (b) follows from (a) by substituting 'A' for '*x*' in (T); steps (f) and (g) hold because their antecedents are provably false according to the (assumed) syntax of the truth-theory; (h) follows from (e), (f), (g), and (T).)

Of course, this method will not work for our language and its infinitely many sentences. But suppose that we fix part of our language as an object language and add a device for forming infinite conjunctions to our metalanguage. Then we can define '*x* is true' by:

x is true if and only if InfConj⟨*x* = '*p*' only if *p*⟩

where 'InfConj⟨x = 'p'only if p⟩' represents an open sentence which is an infinite conjunction of the open sentences of the form:

x = 'p' only if p.

Then by using infinite versions of the derivation given above we can prove a disquotational biconditional for each sentence of our object language.[15]

Notice that the definition of 'is true' in terms of infinite conjunction uses no semantical terms nor any kind of correspondence or epistemic predicates. Thus conceiving of truth as a kind of infinite conjunction need not commit one to either a correspondence or an epistemic conception of truth. Of course, in giving *truth-conditions* for infinite conjunctions we will be forced to talk of truth of some kind. But the same is so when we give truth-conditions for, say, conditionals. Furthermore, one can *state and use rules of inference* for both conditionals and infinite conjunctions without giving truth-conditions for them, and surely simply using infinite conjunctions no more commits one to a view of truth than simply using conditionals does.

Earlier we saw how in languages like ours, truth-predicates substitute for infinite conjunction in inferences whose premisses cite entire theories. We have just seen that giving truth-predicates such a role does not commit one to a correspondence conception of truth—at least so long as simply using infinite conjunctions does not. Thus it is also sometimes useful to think of truth as a form of infinite conjunction. Thinking of truth in this way is to think of a truth-predicate as a kind of logical operator, since infinite conjunction is a logical operator. Thus I will call this way of thinking about truth the *logical conception of truth* and codify it as follows: a truth-theory meets the logical conception of *truth* just in case for each

[15] We can also reverse these definitions. Assume that we already have a truth predicate and disquotational biconditionals for part of our language taken as an object language. Then we can define infinite conjunction as an adjunct to our meta-language, so long as we restrict it to sentences of the object language in question. The definition runs as follows:

InfConj{S: . . . S . . .} if and only if for each S, . . . S . . . only if 'S' is true.

Here 'InfConj{S: . . . S . . .}' represents the infinite conjunction of all sentences S satisfying the condition . . . S . . ..

sentence *S* (within the scope of the theory), it implies a disquotational biconditional for that sentence, that is, a sentence of the form

> *X* is true if and only if *q*,

where '*X*' and '*q*' are schematic letters, respectively for a name of the sentence *S* and the sentence itself (or a translation of it if the sentence be foreign).

To recapitulate the main points of this section: first, we require disquotational biconditionals for truth in order to validate inferences conforming to patterns (1) and (2) presented at the beginning of this section. Secondly, if all we want from a truth-theory is that it yield disquotational biconditionals for truth, then it need only meet the logical conception of truth.

2.3. *Immanent vs. Transcendent Truth*

Conceptions of truth that lead one to build truth-theories covering no sentences beyond one's home language are *immanent*. Conceptions of truth which require one to develop a truth-theory applying beyond one's own language are *transcendent*. Or to put the distinction in terms of truth-theories instead of ways of thinking about truth, a truth-theory is immanent if its scope does not extend beyond its home language, while it is transcendent if it applies to sentences in at least one other language (and strongly transcendent if it applies to sentences in arbitrary languages).

Ordinarily, we think of logical operators, whether finite or infinite, as composed of sentences drawn from one and the same language. Taking this a step further, thinking of our own truth-predicate as a device substituting for infinite conjunctions of our own sentences suggests that our truth predicate might be restricted to our own language. In so doing we would be thinking of truth as immanent.

One reason for taking the immanent approach to truth is that the currently known transcendent truth-theories are based upon problematic theories of reference, meaning, or translation, or else upon controversial epistemic notions that are difficult to apply to our own sentences. Thus, while remaining prepared to embrace transcendent truth if and when it can be placed upon firmer footings, one might for the nonce seek an immanent truth-theory that avoids problematic transcendent foundations. I will call such an approach to truth *weakly immanent*. On the other hand, an immanent truth-theorist might hold that the difficulties with transcendent truth-theories are more fundamental, that at some level talk of transcendent truth or reference does not make sense, and that all proper truth-theories are immanent. I will call such an approach to truth *strongly immanent*. I will argue shortly that a truth-theory answering to an immanent conception is all that mathematical realists need for their realism. This leaves them the option of taking either the strong or weak approach. I am inclined to take the strongly immanent approach to truth, but I will not try to defend that view here.

(Incidentally, it is not possible to classify truth-theories themselves as strongly or weakly immanent, since truth-theories typically do not contain assertions concerning the legitimacy of applying the predicate 'true' to sentences in other languages. Thus it could happen that a person subscribing to the strongly immanent conception of truth develops a truth-theory that can be extended to other, perhaps even arbitrary, languages.)

Adherents of the correspondence or epistemic conceptions normally think of truth as transcendent, since they attempt to characterize truth using language-transcendent reference relations or epistemic properties. But one could, somewhat contrary to the usual spirit of these conceptions, develop an immanent correspondence or epistemic truth-theory by simply not including foreign sentences within its scope.

Instead of developing a truth-theory within a metalanguage containing infinite conjunction as a connective, I prefer the *disquotational* pattern pioneered by Tarski. This method defines truth (for an object language within a containing metalanguage) by using set theory in place of infinite conjunction.[16] Two of its features concern us here. First, this sort of truth-theory implies all the disquotational biconditionals for its object language (and not just the ones for truth). More precisely, it implies each instance of

(Disquot. T): 'p' is true if and only if p;
(Disquot. Sat): x satisfies 'F' if and only if x is F;
(Disquot. Des): 't' designates x if and only if $t = x$,

where both occurrences of 'p' ('F', 't') are supplanted by one and the

[16] Alternatively, one can avoid the set theory by taking 'true' (more precisely, 'satisfies') as a primitive and introducing the usual recursion clauses for it as axioms. Set theory is needed to convert the resulting inductive 'definition' into an explicit one.

same object-language sentence (predicate, name). Second, this sort of truth-theory uses disquotational (or *list-like*) definitions for name and predicate reference. A list-like definition of English name-reference (ENRef) would run along the following lines:

> ENRef(t,x) if and only if
> (t = 'Adam' only if x = Adam) & (t = 'Babs' only if x = Babs) & . . . & (t = 'Zeb' only if x = Zeb),

while one for English one-place predicate-reference (E1PRef) would look like this:

> E1PRef(x,P) if and only if
> (P = 'agile' only if x is agile) & (P = 'burnt' only if x is burnt) & . . . & (P = 'zany' only if x is zany).

Notice that in both cases the lists in question are *finite* conjunctions. Extending these to infinite conjunctions involving compound names and predicates would produce definitions of satisfaction and designation analogous to the one given for truth earlier.

One reason for preferring disquotational truth-theories for working out an immanent, logical conception of truth is that they do not require us to augment our language with controversial logical operators such as infinite conjunction. Another is that the Tarski recursions used in these theories relate the truth-values of compound sentences to those of their components, and set up similar relationships for satisfaction and designation.

The disquotational biconditionals of these theories emphasize their affinities with (the common view of) correspondence truth-theories, while their list-like definitions of reference serve to distinguish them from such theories. Let me elaborate on these similarities and differences more fully.

The schemata (Disquot. T), (Disquot. Sat.), and (Disquot. Des) furnish a very clear sense in which the disquotational approach is compatible with the idea (to quote a characterization of correspondence truth due to Michael Devitt), that 'sentences . . . are true or false in virtue of: (1) their objective structure; (2) the objective referential relations between their parts and reality; (3) the objective nature of that reality'.[17] For even on the disquotational approach, satisfaction and designation are relations between parts of sentences

[17] Devitt (1984), 28.

and objects in the world, and whether or not they obtain is an objective matter of the way the world is. (For instance, whether or not the Sun satisfies 'is a planet' depends upon whether or not the Sun is a planet.) Moreover, even the list-like specifications of name and predicate reference identify word–world relations by virtue of having argument places for terms referring to words and for terms referring to non-linguistic objects. What is more, these specifications agree (or are at least presumed to agree) extensionally over their home language with correspondence accounts of reference.

Despite this, a disquotational theory of truth for English, say, *differs* importantly from a correspondence theory of truth for English.

One obvious difference is that the disquotational theory is an immanent theory that has no pretensions of applying to other languages. In constructing a disquotational truth-theory one's primary aim is to capture the disquotational biconditionals; the defined truth-predicate, the recursion clauses for compound expressions, and the list-like definitions are subservient to this aim. In a language containing infinite conjunction they would be superfluous, since truth, satisfaction, and designation could be defined using infinite list-like definitions. This clearly differentiates thinking about truth as logical operator from thinking about it in correspondence terms. In the latter case the reference (or other 'correspondence') relation necessarily plays a fundamental role, and the theory one obtains purports to be a theory of truth for arbitrary languages.

Another difference between the two approaches is that correspondence theorists hope to give an account of truth and reference suitable for use in general theories of human psychology and linguistic behaviour. One could hardly see this as an aim of the disquotational approach. Tarski's list-like definitions apply only to languages whose vocabularies are finite and fixed and consist of expressions whose extensions are stable. The method for defining reference cannot apply, for instance, to English *qua* evolving language. Nor will the list-like approach support theoretical links between reference relations in different languages. It does not even account for the relation between, say, English one-place predicate-reference and English two-place predicate-reference.[18]

A more subtle but important difference concerns the status of the

[18] Hartry Field has emphasized that the list-like approach failed to be explanatory. He took this to be a drawback of the Tarskian definition. See Field (1972).

disquotational biconditionals on the two theories. In a disquotational truth-theory, the biconditional

'Snow is white' is true if and only if snow is white

is a consequence of the definition of 'true', the syntax of the object language, and the *logic* (or set theory) employed in the metalanguage. On the correspondence approach, the same biconditional is derived from the definition of 'true', the syntax of the object language, and presumably *empirical* hypotheses concerning the transcendent reference relation that replace the list-like definitions of the disquotational approach.[19]

3. REALISM AND IMMANENT TRUTH

Mathematical realists, along with many other realists, need to talk of truth, because they want to assert the truth of various theories. If these theories could be condensed into a single sentence *S*, then mathematical realists could meet conditions (2) and (3) by simply asserting that *S* and that *S* whether or not we have proved it. (They would have to leave it understood that they were committing themselves to all of *S*'s logical consequences too.) It happens that by using second-order logic one can condense number theory, analysis and Zermelo–Fraenkel set theory into a single sentence *S*. But not every mathematical realist is a fan of second-order logic, and no interesting branch of mathematics can be condensed to a single first-order sentence unless one supplements the mathematics in question with predicates foreign to it. So most mathematical realists require a truth-theory. Yet they require no more than an immanent, logical truth-theory. In particular, an immanent, disquotational truth-theory of the type I prefer will suit them fine.

Immanent realism is realism with truth conceived immanently. I take myself to be an immanent mathematical realist, and I have just claimed that immanent realism is all the realism mathematical realists need defend. However, due to its limited aims, the immanent conception of truth and the realism defined in terms of it may appear to be much weaker than that found in traditional metaphysics. In discussing Quine's views on truth and reference Hilary

[19] This point is explained very well in Etchemendy (1988).

Putnam once had this opinion. Calling Quine's views 'immanent', he described them as follows:

> [For Quine] to say *'Snow is white' is true* is to reaffirm 'Snow is white' and not to ascribe a mysterious property called 'truth' to 'Snow is white.' Similarly, we might say that for Quine reference is 'immanent reference'—to say *'Cat' refers to cats* is to say only that cats are cats, and not to say that a mysterious relation called 'reference' obtains between the word 'cat' and cats. Any definition of reference that yields the truisms '"Cat" refers to cats,' '"Electron" refers to electrons,' etc. will do. We do not have to *first* 'put the words in correspondence with objects' and *then* utter these statements to declare which objects our words correspond to.[20]

Then he continued by criticizing this view of truth as inimical to realism:

> When I say that I am trying to decide whether what you have said is true, then, in Quine's view, all I mean is that I am making up my mind whether to 'assent.' But this is to give up what is right in realism. The deep problem is how to keep the idea that statements are true or false, that language is not merely noise and scribbling and 'subvocalization,' without being driven to postulate mysterious relations of correspondence. Quine's view is not the cure for metaphysical realism but the opposite pole of the same disease.[21]

Now if you say, 'No uncountable set of reals has a cardinality smaller than the continuum,' and I am trying to decide whether what you say is true, then I am trying to decide whether no uncountable set of reals has a cardinality smaller than the continuum. As we saw in the last section, not only do truth-theories answering to the logical conception readily imply that this is what I am trying to decide, they are among the few truth-theories that do so with ease. It is unfortunate that Putnam seems to be sufficiently misled by his way of characterizing disquotational approaches ('to say *"Snow is white" is true* is to reaffirm "Snow is white"'), to suggest that when I am trying to decide whether what you said is true, I am simply trying to decide whether to endorse your utterance. Of course, if Quine's view

[20] Putnam (1984), 12.

Quine also speaks of the immanent–transcendent distinction in ontology in Quine (1969a) and of immanent/transcendent grammar and truth in Quine (1970). In Quine (1981c) he says 'Immanent truth, a la Tarski, is the only truth I recognize' (p.180). It is not clear whether Quine agrees with my characterization since Tarski sometimes says that object language sentences can be disquoted by *translations* in the metalanguage.

[21] Putnam (1984), 13

amounted to this, it would be wrong: reasons having nothing to do with the truth of the continuum hypothesis might move me to endorse your utterance.

Putnam rightly points out that the disquotational approach to truth says nothing about *how* language relates to reality and that it fails to explain the nature of the relationship. It does not fail to do this by being immanent, since immanent correspondence theories are possible, but by being a form of the logical approach. Now to be a realist about mathematical objects, theoretical entities in science, or possible worlds is to adopt a metaphysical view of these entities, and not a semantic or epistemic theory concerning them. In their metaphysical modes, realists need to speak of truth only to assert theories about the entities they recognize. Thus it makes sense that they require only an immanent, logical truth-theory. To support this claim further I will examine other commonly voiced demands on realism and truth in order to show that either they are not relevant to the debate between mathematical realists and anti-realists or else they can be accommodated (to a large extent) within an immanent approach to truth conceived of as a logical operator.[22]

3.1. *Demands from Metaphysics*

Some philosophers believe that realism is committed to the view that truth depends upon features of the world 'out there'. If in the mathematical case this just means that our theories are true or false independently of our proofs and constructions, then it can be accommodated using an immanent, logical conception of truth. For, if we had infinite conjunctions, we could formulate versions of this thesis for specific sentences without talking of truth, and then explicate the general thesis as the infinite conjunction of these specific claims. In other words, we could form the infinite conjunction of sentences such as:

> There are infinitely many twin primes (or not) independently of our proving so (or otherwise); the cardinality of the continuum is smaller than the second aleph (or not) independently of our proving so (or otherwise); etc.

[22] The idea that realists do not require correspondence truth is not new. See Devitt (1984). I differ from Devitt in taking 'realism' to apply to a broader collection of doctrines than he does and in seeing truth as essential to some very strong forms of realism.

In our language, we say

> Every mathematical sentence is true or not independently of our proving or refuting it.

But since here we need only use the predicate 'true' to substitute for an infinite conjunction, we need presuppose no more than a logical conception of truth.

On the other hand, the thesis might be taken to mean that a (mathematical) sentence is true just in case some non-linguistic, non-epistemic condition associated with the sentence obtains, that is, that truth is a matter of the way the world is. Again we can explicate this idea for a specific sentence, say, 'The Earth is a planet.' For this is true or false just in case a specific worldly condition obtains, namely, the condition of the Earth being a planet. But since this obtains if and only if the Earth is a planet, the biconditional

> 'The Earth is a planet' is true if and only if the Earth is a planet

serves to explicate the idea that the truth of this specific sentence depends upon the way the world is. Then, as in the last example, we can explicate the general claim via the infinite conjunction of all sentences of the form:

> '*p*' is true if and only if *p*.

However, in order to express this conjunction using a truth-predicate we must take a different approach from that adopted in the last example. We cannot rewrite the infinite conjunction as:

> For any sentence *S*, *S* is true just in case ——

because this fails to indicate the connection between the sentence variable and the blank. Nor can we rewrite it as:

> For any sentence *S*, *S* is true just in case *S*,

because this will cause '*S*' to function both as a variable ranging over sentences and as a schematic sentence letter. What we need is something of the form:

> For any sentence *S*, *S* is true if and only if *G*(*S*),

where '*G*(*S*)' is an open sentence with the free variable '*S*'. Now the Tarskian definition of 'true' is a biconditional with '*S* is true' on the left side and the Tarski *definiens* as its '*G*(*S*)'. Furthermore, it implies all the specific conjuncts of the infinite conjunction we need to replace. So we can use it as our '*G*(*S*)'.

If we are willing to appeal to translation, then we can even apply the methods of the last two examples to translated foreign sentences as well. To say, for example, that such a sentence is true in virtue of features of the world is to affirm a biconditional whose left side attributes truth to the sentence and whose right side is the translation of the sentence. (Of course, immanent truth-theorists will not pursue this application.)

On the other hand, some metaphysicians may want to advocate a stronger view than we have captured here. For example, they may hold that there are *truths* (in the sense of facts) which obtain independently of us, our languages and our theories, and which may not be formulated in any language that we currently possess, or perhaps in any language at all. The full articulation of this philosophical doctrine probably requires substantial commitments concerning both the nature of truth and truth-bearers (one may need to postulate propositions that are not expressed in any language), and it is unlikely that its proponents will be satisfied with truth conceived immanently or as a logical operator. For such metaphysicians, immanent mathematical realism is not a true realism. Rather than attempt to refute this view directly, which would be no easy task since it has not been fully articulated, I will simply point out that contemporary mathematical realists have not maintained such a strong metaphysical thesis nor have anti-realists attacked them for holding such a view. Instead the debate between them has concerned the truth and ontologies of extant mathematical theories.[23]

3.2. *Demands from Philosophy of Science*

History teaches us that sooner or later we will be forced to abandon or qualify even our most successful scientific theories. Thus many scientific realists maintain only that contemporary science is approximately true rather than categorically true. Some, fearing that even

[23] This is confirmed by the definitions of realism Field and Maddy use when, respectively, rejecting and defending it. See Field (1989) and Maddy (1990). Some mathematical realists have committed themselves to a stronger view than immanent realism. Recently Bernard Linsky and Edward Zalta have done so in propounding their platonized naturalism. See Linsky and Zalta (1995).

this claim may eventually succumb to empirical evidence, identify their scientific realism with the thesis that science aims to produce true theories.[24] These philosophers require a transcendent conception of truth, in so far as they think of the truth as independent of any specific language or an approximately true theory as one which, for example, more or less corresponds to extralinguistic facts that may or may not be formulated in the language of the theory in question.

Philosophers of mathematics have not proposed versions of mathematical realism of this ilk, because they tend to overlook the fallibility of mathematics. The history of mathematics is much less a series of failed or flawed theories than the history of empirical science. Yet the historical considerations motivating philosophers of science arise in the history of mathematics too: the inconsistency of 'naïve' set theory and the ill-defined concepts of the calculus are just the two best-known cases. So some mathematical realists may desire a richer sense of truth than I have provided.

One way to avoid taking such a step is to speak of our current mathematics as consisting mostly of true sentences (or as containing a certain proportion of truths). An even cleaner approach would be to select, perhaps in consultation with one's anti-realist opponents, a certain core set of mathematical assertions and stake one's realism on these. This would permit realists to assert that mathematics is basically true while separating the fate of realism from fluctuations in the more speculative parts of mathematics. Moreover, these qualified versions of mathematical realism would still allow their proponents to engage in the traditional debates with anti-realists concerning the reality of mathematical objects and our ability to know them.[25]

Realists who accept these recommendations can do so within the confines of immanent, logical truth, so long as the mathematics (or parts of it) they endorse is affirmed in their home language. For whether they assert that mathematics is entirely true, basically true, or mostly true, they can be construed as asserting the (possibly infinite) conjunction of the members of some set of sentences.[26]

[24] See the editor's introduction to Leplin (1984)

[25] Cf. Devitt (1984), 22–3, where a similar point is argued concerning scientific realism.

[26] Even those realists who feel compelled to hold that mathematics is only approximately true, need not be committed to correspondence truth. The theories of

3.3. *Demands from the Theory of Cognition*

One sort of argument that one finds in favour of correspondence truth-theories is that they are necessary to explain why people with true beliefs are more likely to succeed and thrive. The idea seems to be that someone who has true beliefs will, on the correspondence conception, be in a better position to interact with the environment than someone who has false beliefs. Now if this idea has any useful content it comes down to the view that people who have true beliefs will be in a better position to perform successful actions. But we need not assume a correspondence theory to explain why this is so. For beliefs to guide actions successfully they need only imply all the true observation statements (and no false ones) that are relevant to the actions in question. We need no notion of truth richer than a logical one to state this point. Arguing for it is another matter. I am not sure that we can argue on either the logical or the correspondence conceptions of truth that true non-observational beliefs are more likely than false ones to imply true observation statements. (We could argue this using an epistemic theory that entailed that a true theory must have all the necessary implications.) But since a theory that implies at least one false observation statement is itself false, we can see some connection between having false beliefs and failure to thrive. Yet it is also a connection the logical conception of truth can establish.[27]

On the other hand, a cognitive scientist might want to affirm that any species as intelligent as ours will develop some systematic body of mathematical truths. This would require a richer conception of truth than an immanent one, though it is hard to tell what form it might take. However, the question of whether other intelligent

approximate truth developed by Ilkka Niiniluoto and Thomas Weston do not define approximate truth in terms of correspondence, but rather in terms of measures of 'distance' from an unspecified true theory. The latter notion of truth can be taken as a form of the logical conception of truth. See Niiniluoto (1987) and Weston (1992)

[27] The logical conception implies that a theory that implies something false is itself false.

To guide our actions, it is not enough that a theory implies observation sentences, it must imply ones that are relevant to them. So don't we need a correspondence theory to identify the relevant observation statements? No, since they can be taken to be tied to experience as wholes. (Cf. Quine's treatment of observation sentences in Quine (1990).) This would identify the relevant beliefs by reference to the experiences associated with them, thereby avoiding the correspondence theorist's approach of correlating sentential components to parts of reality.

beings will develop a (true) mathematics has not been an issue in realist/anti-realist debates about mathematics, and I see no reason why it should become one.

3.4. *Demands from Philosophy of Language*

Many philosophers of language are unlikely to be satisfied with either an immanent or a logical conception of truth. To some the transcendence of truth is fundamental both to our our common-sense notion of truth and to theoretical semantics. Some also see the biconditionals for truth as empirical hypotheses, which are either the essential ingredients of a theory of meaning for a language or an important component of one. 'How', they might ask, 'can a proper truth-theory yield such truths as definitional transforms of logical truths?'

Let us take up this second problem first, since we can dispose of it quickly. Consider the geometric sentence, 'The sum of the angles of a triangle is 180 degrees.' If we construe this as a claim about Euclidean space(s), then it is true by definition; for it is a logical consequence of the definition of a Euclidean space that the sum of the angles of its triangles are 180 degrees. On the other hand, taken as a claim about physical space, it is empirically false; because space is not Euclidean. Similarly, the disquotational biconditionals for a language L are true by virtue of the Tarskian truth-definition for L, but this does not preclude its remaining a non-logical question as to whether a given target population speaks the language L. It is also important to remember that our measuring, say, the angles of triangles to determine whether our space is Euclidean depends upon our interpreting the Euclidean notions of line, angle and triangle in physical terms. Similarly, determining whether a given population speaks a given language will depend upon our interpreting that language in terms we understand.[28]

I am sympathetic to the first objection to immanent truth. Although recognizing the truth of some untranslated body of theorems is not essential to realism, I can see that some realists might

[28] For the sake of my exposition I am assuming that whether physical space is Euclidean is an empirical question, despite the hot debates among philosophers of physics on this issue. The issue of whether it is an empirical question as to which language a population speaks is no less complex than the corresponding issue about space. That is why I have used the term 'non-logical' instead of 'empirical'.

want the freedom to so. I think they can attain this within an immanent framework. Here is the basic idea. First notice that since most of us philosophers are not particularly fluent in even mathematical English, the difficulty of claiming that a certain piece of mathematics is true can arise even in the case of our native tongue. We have every reason to believe that the theorems announced in the latest issue of the *Journal of Symbolic Logic* are true but usually we only partially understand them at best. Now just as our own idiolects fall short of mathematical English, they also fall short of mathematical and scientific German, French, and other foreign technical dialects. Yet since we usually think of our language as including its technical extensions, we could just as well extend the same courtesy not only to foreign technical languages but also to all human language—considered as a polyglot.

I am suggesting that we think of our truth-predicate as applying immanently to the human polyglot. But in order to keep to the immanent, logical conception we have used so far we must have the disquotational biconditionals for non-English expressions follow the same pattern as the ones for English. More exactly, when disquoting a foreign sentence or name or predicate we simply put the quoted part of speech in the appropriate place of the disquotational schema just as we have done so far for the English examples.[29] Thus the biconditional for 'un et un font deux' is

> 'un et un font deux' is true if and only if un et un font deux,

instead of

> 'un et un font deux' is true if only if $1 + 1 = 2$.

We can recover the latter if we know that un et un font deux if and only if $1 + 1 = 2$. Of course, this depends upon allowing sentences which mix languages, but this is just an extension of our usual practice of absorbing certain foreign expressions into English.

On transcendent approaches to truth, however, we have disquotational biconditionals of a transcendent truth-theory connecting quoted foreign sentences with unquoted English counterparts. Thus the biconditional

[29] I am supposing that these languages have the same logical forms. Dealing with languages that do not (if there is a fact of the matter here) is a problem for any theory of truth or reference, immanent or transcendent, that aims to disquote foreign expressions.

'un et un font deux' is true if and only if $1 + 1 = 2$

is part of the truth-theory itself, and not a consequence of the truth-theory plus a mixed biconditional not in the truth-theory. Furthermore, depending on which transcendent approach is in play, the statement that a body of foreign statements are true means either (a) that their English translations are disquotationally true, or (b) that they could be translated as such English sentences, or (c) that their truth-conditions are satisfied. Each of these alternatives is controversial, since they respectively appeal to the notions of translation, possible translations, and truth-conditions.

I think of my proposal as a form of immanent truth since I think of the human polyglot as one big language. Granted this assumption (which seems less controversial than the just-mentioned assumptions which transcendent theories require), we can gain some of the benefits of a transcendent theory without its costs.

On the other hand, adhering to this extended immanent conception of truth will not permit us to discuss whether, say, any rational being will know some mathematical truths or other issues about mathematical truths not expressed in any human language. Here I will just say what I have said before: issues such as these have not arisen in the course of the debates between mathematical realists and anti-realists, and I can think of no reason why mathematical realists would be required to take a stand on them.

4. SOME CONCLUDING REMARKS

Some philosophers have remarked to me that immanent realism is a platitude; still others that it converts important anti-realist positions into definitional falsehoods. 'Consider, for example,' the latter say, 'the view that mathematical sentences are mere contentless pieces in a game. Using a disquotational truth-theory for mathematical sentences we can prove, contrary to this thesis, that any given mathematical sentence has a truth-value.'[30] However, the problem is not due to the immanent, logical conception of truth. The anti-realists in question claim that certain sentences should not be understood as ordinary declarative sentences and thus are no more true or false

[30] Sample proof: By excluded middle, $1 < 2$ or not $1 < 2$. Then using the disquotational biconditional for '$1 < 2$', we get '$1 < 2$' is true or '$1 < 2$' is false (not true).

than imperatives or questions. They deny that *any* truth-theory applies to the sentences in question. But this does not mean that they cannot accept disquotational truth where they think talk of truth applies. The debate between them and realists still concerns the traditional issue between them, that is, the status of sentences formulated in the vocabulary of mathematics.

I am not sure of how to interpret the idea that immanent realism is a platitude.[31] I suppose that it means that no one is going to deny that, say, there are infinitely many prime numbers or other mathematical truths, but they may well deny that such truths are to be understood as committing us to the existence of abstract objects. If this is what the remark means, then it is clearly wrong. Many anti-realists do deny standard mathematical claims. Hartry Field denies *all* mathematical existence claims including the one cited, while constructivists deny numerous claims of standard non-constructive mathematics. Furthermore, although there are anti-realists who affirm standard mathematical truths, they contradict immanent realism by denying that mathematical objects exist or else that they exist independently of our proofs and constructions.

An opposite worry is that immanent realism can be refuted by discovering false mathematical theorems, and that a philosophical thesis should not be refuted by evidence of this sort. In Section 3.2 I discussed ways of qualifying the truth claims of scientific and mathematical realism in order to avoid refutations of this sort. Let me add that I view the debates between realists and anti-realists in these areas as concerned with whether we ought to believe the claims of science and mathematics. Part of the reason that realists think that we should is the strong evidence, scientific and mathematical, for these claims. Of course, if we learned that the evidence did not support a given branch of science or mathematics, we would have little reason for accepting it and thus little reason for taking a realist stance towards it. In the period immediately after the discovery of the paradoxes, philosophers and mathematicians probably had good reasons for not being realists about sets, and today there is still good reason to be cautious in one's realism about the contemporary extensions of set theory.

[31] The suggestion was put to me by an anonymous referee.

3

The Case for Mathematical Realism

1. THE PRIMA FACIE CASE FOR REALISM

As intelligent persons we take mathematics seriously. When we are told that the Greeks discovered lengths that cannot be expressed as fractions of any unit length, we presume that we hear the truth, just as firmly as when we hear that animals roamed the Earth long before humans existed. The practice and language of mathematics, its exalted place in our intellectual life, and its enormous technological successes all promote the idea that mathematics is a factual science with it own subject-matter.

Let me elaborate a bit on why this is so. I will start with the point about the language of mathematics. Notice that the apparent grammatical and logical forms of mathematical existence-claims are the same as those of more mundane existence-claims. This can be brought out dramatically by considering a sentence which mixes these claims. On a straightforward reading the sentence

> The solutions to some of the problems involved numbers exceeding the capacities of some pocket calculators

contains to two existential claims: first, that numbers exist which solve certain problems, and second, that there are pocket calculators that cannot handle these numbers. Consequently, anyone who denies the straightforward reading of the claim about numbers owes us an account of the type of claim, if any, it does make. But no uncontroversial account is at hand, and a history of disappointing attempts hardly makes the prospects of developing one encouraging.[1]

[1] It might be easier to interpret 'there is a' followed by a mathematical term as a type of existence claim that somehow lacks the existential force of other existence claims. But this would still leave us with the mystery of how mathematical terms can

Not only do we presume that mathematicians mean what they say—that their existential claims are to be read straightforwardly—we also presume that their pronouncements result from an earnest and well-disciplined search for the truth. Mathematics has all the trappings of a science—refereed journals, results that can be replicated and independently checked, and so on. Furthermore, natural scientists incorporate mathematical results in their research in much the same way that they incorporate the results of their own and other branches of science. None of them seem to think that mathematics is just make-believe.

Finally, mathematics works. Without numbering, measuring, calculating, and mathematical modelling we would never have developed even our pre-computer technology. And with the advent of the computer mathematics is destined to play an increasingly important role in our technology. What is more, mathematics provides more than just the techniques of numbering, measuring, and calculating; it also tells us some of the things we can and cannot do with our mathematics-based technology. The theory of algorithms, for example, tells us that there are certain problems that computers cannot solve, while the theory of computational complexity tells us that certain kinds of algorithms are generally more efficient than others. Only physics, chemistry, and biology can come close to claiming the benefits that mathematics has given us, and even they could not do without mathematics. Because of this it strains the imagination to think that mathematics is an elaborate game, fable, or art form that has just happened to prove useful.

In fact, if mathematics were just a game or an art we could not explain its usefulness, because we do not *use* them in the way we use mathematics. They may teach, entertain or enlighten, but they do not supply premisses for scientific and practical inferences. Mathematics does. This still leaves the view that mathematics is a fable. One might also employ the assertions of a fable as one's premisses, but why would anyone consciously do so?[2]

undo the existential force of a 'there is' while other terms cannot. Simply using one universe of discourse for mathematical objects and another for others will not work, for 'there is a mathematical object . . .' will still assert the *existence* of a member of the mathematical universe. Moreover, using multiple universes need not be salutary, since counting collections containing both mathematical and non-mathematical objects will then require some fancy logical moves.

[2] Actually, the idea that mathematics is a fable is not so easily dismissed. I will deal with fictionalist approaches more carefully in Chapter 4.

The pervasive use of mathematics in science and technology is probably the chief reason that mathematical realism has survived so many challenges. Because many contemporary philosophers of mathematics regard the argument from applications as the strongest case for realism, it has been the focus of many of the recent debates between realists and anti-realists. In view of this, I will devote the rest of this chapter to a detailed analysis of this argument.

2. THE QUINE–PUTNAM VIEW OF APPLIED MATHEMATICS

We owe the realist's account of applied mathematics to W. V. Quine and Hilary Putnam. They maintain that applying mathematics is an *indispensable* part of scientific practice. First, mathematical language is needed to give scientists an apparatus for representing empirical findings. Referring to mathematical objects allows scientists to introduce such concepts as acceleration and state vector into physics, random mating and allelic frequency into genetics, expected utility and welfare function into economics. Second, mathematical laws are required for inferring non-mathematical conclusions from those non-mathematical assumptions that have been formulated with the help of a mathematical vocabulary. Eliminating mathematics would thus drastically curtail science.

Quine and Putnam emphasize that in using mathematical terminology and premisses scientists are not merely using the formalism of mathematics, *they are also presupposing the existence of the mathematical objects and the truth of the mathematical principles.* For example, the concepts of acceleration, velocity, state vector, random mating, allelic frequency, expected utility, and welfare function are all defined in terms of the real numbers. If they did not exist, then these concepts would be no more well-defined than that of the sum of a divergent infinite series, and all generalizations framed in terms of them would be vacuously true. Furthermore, if standard mathematics were not true, we would have no reason to believe in the soundness of the mathematical deductions scientists use, or the correctness of their calculations, or the cogency of their statistical reasoning.

In the rest of this chapter it will important for us to remember that presupposing the existence of mathematical objects and the truth of large portions of standard mathematics is by no means restricted to the application of mathematics to theories that scientists regard as true. Even when they develop a purely speculative theory or a highly idealized model, scientists presuppose the truth of the mathematics they use. For the models will not have the properties they are supposed to have unless the background mathematics holds.

To illustrate this point, let us consider Newton's account of the orbits of the planets. He calculated the shape of the orbit of a single planet, subject to no other gravitational forces, travelling about a fixed star. He knew that no such planets exist, but he also believed that there are mathematical facts concerning their orbit. In deducing the shape of such orbits, he presumably took for granted the mathematical principles he used. For the soundness of his deduction depended upon their truth. Furthermore, in using his (mathematical) model to explain the orbits of actual planets, he presumably took its mathematics to be true. For he explained the orbits of planets in our solar system by saying that they approximate the behaviour of an isolated system consisting of a single planet orbiting around a single star. For this explanation to work it must be true that the type of isolated system (Newtonian model) has the mathematical properties Newton attributed to it.

It is also important to note that some fairly sophisticated mathematics is at work in fairly mundane areas of science. Contemporary biology, economics, psychology, and sociology are bursting with statistics. Here scientists use real analysis to compute statistical measures of data taken from finite populations of observable objects. In claiming, for example, that some data are normally distributed they presuppose that real numbers defining the data curve exist; and in computing means and standard deviations they presuppose the truth of a variety of mathematical equations. Mathematics functions for them as a background framework within which they may formulate laws and theories, build models, and carry out inferences.

3. INDISPENSABILITY ARGUMENTS FOR MATHEMATICAL REALISM

The Quine–Putnam account of applied mathematics can be summarized in the following indispensability thesis:

> *Indispensability*: Referring to mathematical objects and invoking mathematical principles is indispensable to the practice of natural science.

Now nothing about the reality of mathematical objects follows from this alone, but suppose that, following Quine's lead, we adopt the next two theses:

> *Confirmational Holism*: The observational evidence for a scientific theory bears upon the theoretical apparatus as a whole rather than upon individual component hypotheses.
>
> *Naturalism*: Natural science is our ultimate arbiter of truth and existence.

Then we can construct a so-called indispensability argument for mathematical realism along these lines: mathematics is an indispensable component of natural science; so, by holism, whatever evidence we have for science is just as much evidence for the mathematical objects and mathematical principles it presupposes as it is for the rest of its theoretical apparatus; whence, by naturalism, this mathematics is true, and the existence of mathematical objects is as well-grounded as that of the other entities posited by science. For future reference let us call this the Holism–Naturalism (H–N) indispensability argument.[3]

Now one way to criticize this argument is to challenge its premisses. In the next chapter, I will look at a number of attempts to undermine the indispensability premiss through showing that we need not presuppose the (literal) truth of mathematics in natural science. Here I want to take up different sorts of challenge recently propounded by Penelope Maddy and Elliott Sober.[4] Neither thinks that we can count on science to provide evidence for the truth of mathematics. Sober claims that scientific testing fails to confirm the mathematics used in science. In Chapter 7 I will take up Sober's objection, which is really an objection to confirmational holism itself. Maddy's criticism is based upon observing that much of the

[3] This is sometimes called the *Quine–Putnam indispensability argument*. However, because Quine and Putnam never formulated their arguments as explicitly as I have, it is not clear whether they intended this argument or the pragmatic argument that I formulate below. The view that in applying mathematics we presuppose its truth can be traced to Frege. See Resnik (1980), 62–3.

[4] See Maddy (1992) and Sober (1993).

mathematics used in science occurs in predictively useful theories that scientists openly acknowledge as false, and that many scientists distinguish between parts of a theory that they regard as true and parts that they currently regard as merely instrumentally useful. She argues that this raises the possibility that the confirmation coming from membership in scientific theories that are accepted as true covers too little mathematics to be of comfort to mathematical realists. In short, too much of mathematized science may fall outside the scope of the H–N indispensability argument's naturalism premiss. Here is how she puts it in one passage:

> If we remain true to our naturalist principles we must allow a distinction to be drawn between parts of a theory that are true and parts that are merely useful. We must even allow that the merely useful parts might in fact be indispensable, in the sense that no equally good theory of the same phenomena does without them. Granting all this, the indispensability of mathematics in well-confirmed scientific theories no longer serves to establish its truth.[5]

However, we need not trouble ourselves with Maddy's worry. For whatever attitude scientists take towards their own theories, they cannot consistently regard the mathematics they use as merely of instrumental value. As we saw in the Newton example, even when applying it to idealizations or theories they know are wrong, they use it in a way that commits them to its truth.[6]

Reflecting on this leads one to another indispensability argument—one that is not subject to the objections Maddy and Sober raise and that supports mathematical realism independently of scientific realism. This argument, which I will call the *pragmatic indispensability* argument, runs as follows:

(1) In stating its laws and conducting its derivations science assumes the existence of many mathematical objects and the truth of much mathematics.

(2) These assumptions are indispensable to the pursuit of science; moreover, many of the important conclusions drawn from

[5] Maddy (1992), 281. Maddy also objects to holism on the ground that it gives the wrong picture our evidence for mathematics. I will discuss this issue in Chapter 7.

[6] I tried to argue this in general and nontechnical terms in the last section. In Resnik (1992a) I argue the point in a more detailed way by analysing a piece of mathematical biology.

and within science could not be drawn without taking mathematical claims to be true.

(3) So we are justified in drawing conclusions from and within science only if we are justified in taking the mathematics used in science to be true.

Notice that, unlike the earlier H–N indispensability argument, this one does not presuppose that our best scientific theories are true or even that they are well-supported. It applies whenever science presupposes the truth of some mathematics. Thus, as we noted earlier, it applies even to the mathematics contained in empirically falsified scientific theories, such as Newtonian physics, which we still use in many scientific contexts, and to the mathematics used in constructing idealized scientific models, such as an example Maddy cites of water waves in an infinitely deep ocean. Furthermore, the argument, at least as it stands, contains no claim that the evidence for science is also evidence for mathematics. It has the fairly limited aim of defending mathematical realism by pointing out that any philosophy of mathematics that does not recognize the truth of classical mathematics must then face the apparently very difficult problem of explaining how mathematics, on their view of it, can be used in science.[7]

Still (3) does invite the worry that we would not be justified in using mathematics in science unless we had prior evidence of its truth. If by science we mean disciplines such as physics, chemistry, and biology, then I think we do have quite a bit of independent evidence for mathematics. We need not turn to the a priori either. For the evidential relations between experience and the various branches of mathematics are not very unlike those between experience and the

[7] Despite its modest aims the pragmatic indispensability argument effectively refutes several traditional philosophies of mathematics, such as deductivism (if, thenism), formalism, Quine–Goodman nominalism, and intuitionism. The following quote from Putnam indicates that he and Quine may have intended the indispensability argument this way too:

> So far I have been developing an argument for realism along roughly the following lines: quantification over mathematical entities is indispensable for science, both formal and physical: therefore we should accept such quantification; but this commits us to accepting the existence of the mathematical entities in question. This type of argument stems, of course, from Quine, who has for years stressed both the indispensability of quantification over mathematical entities and the intellectual dishonesty of denying the existence of what one daily presupposes. (Putnam (1971), 57.)

various branches of science. Practice with counting, measuring, surveying, and carpentry suggested and confirmed the elementary rules of practical arithmetic and geometry long before they were elevated to the status of inviolable laws and codified into mathematical systems. Moreover, many of the basic techniques of calculus can be checked geometrically or arithmetically, which in turn helps support its generalizations to real, complex, and functional analysis.[8]

We can also argue for the truth of mathematics on pragmatic grounds by coupling the pragmatic indispensability argument with one to the effect that we are justified in using mathematics in science, because we know of no other way of obtaining its explanatory, predictive, and technological fruits. Here is a formulation of such an argument:

(4) We are justified in using science to explain and predict.
(5) The only way we know of using science thus involves drawing conclusions from and within it.
(6) So, by (3) above, we are justified in taking this mathematics to be true.

Notice that this argument is similar to the H–N argument except that instead of claiming that the evidence for science (one body of statements) is also evidence for its mathematical components (another body of statements) it claims that the justification for doing science (one act) also justifies our accepting as true such mathematics as science uses (another act).[9]

Since much standard mathematics is used in science, the indispensability arguments support realism about many parts of mathematics. Yet as Maddy and others have noted, indispensability arguments fail to cover the more theoretical and speculative branches of mathematics. In Part Two I will argue that other grounds can justify a realist attitude towards branches of mathematics that do not seem to be indispensable to science.

[8] I will argue for the empirical nature of mathematics further in Part Two.

[9] It is likely that our ultimate justification for using science to explain and predict is that doing so appears to promote our theoretical and practical interests better than any other method we know. If so, then our justification for accepting the mathematics used in science also has this character. But isn't this what one would expect from a pragmatic argument?

4. INDISPENSABILITY AND FICTIONALISM ABOUT SCIENCE

Using mathematics to do science does not force one to presuppose the truth of the individual scientific hypotheses themselves. For scientists can employ mathematics to establish *mathematical* claims of the form 'in a (possibly idealized) physical situation of type I, law L holds', even when they do not believe that situations of that type actually obtain. Using such mathematical results they can describe features of theoretical models of situations of type I. In turn they can use these models to make calculations of values or ranges of values of the quantities that interest them. As we noted above, Newton used idealized models of the solar system to calculate orbits for the planets. Once scientists have made their calculations they decide whether their models are good enough for their purposes by comparing the computed values with the data they have (or think they would have) independently obtained. This decision itself often involves mathematical techniques.[10]

Although using mathematics in science does not force scientists to regard scientific hypotheses as true, according to the indispensability thesis, it does commit them to the truth of the mathematics they use. *Even fictionalists about theoretical entities may be committed to the truth of the mathematics used in science.*[11]

It is worth illustrating this by considering the views of Bas van Fraassen, the most prominent anti-realist in the philosophy of science. He regards the entire language of science as both meaningful and truth-valued, but he also holds that we ought not to believe that the theoretical part of science is anything but empirically adequate. Now van Fraassen defines *empirical adequacy* in mathematical terms: a theory is empirically adequate 'if it has some model such that all appearances are isomorphic to empirical substructures of that model', where the appearances are 'the structures which can

[10] Henry Kyburg has described the sort of statistical reasoning they use. One of the purposes for building such models may be to determine whether the theoretical assumptions included in the model are true. See Kyburg (1984).

[11] This remark does not apply to those who hold that theoretical science is a *meaningless* instrument, because, on their view, no inferences of any kind take place within theoretical science. However, explaining how we infer observation statements from features of such an instrument would almost certainly commit them to some mathematics.

be described in experimental and measurement reports'.[12] Thus in claiming that a theory is empirically adequate one asserts the existence of mathematical objects (i.e. models, structures, isomorphisms).

Van Fraassen is no fan of mathematical objects. Here is what he says about them:

> I am a nominalist . . . Yet I do not for a moment think science should eschew the use of mathematics. I have not worked out a nominalist philosophy of mathematics—my trying has not carried me that far. Yet I am clear that it would have to be a fictionalist account, legitimating the use of mathematics and all its intratheoretic distinctions in the course of that use, unaffected by disbelief in the entities mathematical statements purport to be about.[13]

Why does he not claim that it is best to believe that mathematics, like theoretical science, is merely empirically adequate? Perhaps, because he realizes that as his view now stands doing so would still commit him to mathematical objects. To succeed, his fictionalist approach to mathematics must show that the mathematics is not an indispensable component of science proper or of scientific methodology.

5. CONCLUSION

We have seen that the practice of mathematics and its use in science provide a strong prima facie case for mathematical realism. Now scientists and mathematicians talk little of truth, and when they do it is to make simple claims that such and such follows from the truth of so and so theory or that such and such falsifies so and so theory. In short, they use truth in inferences conforming to patterns (1) and (2) of the previous chapter. Probably the only exception to this occurs within mathematical logic where logicians are often concerned to show that various axioms are true under various interpretations.[14] Here they are not using a conception of truth *simpliciter* but rather a mathematical notion of *truth under an interpretation*. So nothing in the practice of science seems to presuppose the truth of mathem-

[12] van Fraassen (1980), 64.

[13] van Fraassen (1985), 303.

[14] Another exception may be philosophers who talk of truth when doing 'cognitive science' or its philosophy. Although these philosophers frequently speak of correspondence truth, they do not appeal to mathematical truths.

atics in anything stronger than an immanent, logical sense. Consequently, the case for realism drawn from the practice of science and mathematics is a case for immanent mathematical realism.

It would be nice if my case were airtight, and I could stop here. Unfortunately, a number of philosophers have argued that scientists could use other, less philosophically suspect, formalisms in place of standard mathematics (interpreted at face value). In short, contrary to my assumption, they hold that to do science it is not necessary to accept standard mathematics (as standardly interpreted). The next chapter will be devoted to examining the prospects for anti-realist alternatives to standard mathematics.

4

Recent Attempts at Blunting the Indispensability Thesis

The indispensability thesis is a premiss of both the holism–naturalism and the pragmatic indispensability arguments. This chapter discusses several important recent anti-realist programmes which either have been explicitly aimed at refuting the indispensability thesis or can be interpreted as directed against it. Recall that the thesis consists of two sub-theses: (1) that using mathematical terms and assertions is an indispensable part of scientific practice, and (2) that this practice commits science to mathematical objects and truths. This gives anti-realists several options: they can try to show that the mathematical formalism is not necessary for doing science, or that using this formalism in science need not commit one to mathematical objects and truths, or finally that a supposed commitment to mathematical existence and truth can be understood in anti-realist terms. Contemporary philosophers of mathematics have pursued each of these options both separately and in various combinations.

In criticizing these programmes I will argue that it is not evident that any of them succeeds in demonstrating that science can avoid a realist commitment to mathematical truths and mathematical objects. I will also contend that even if these programmes succeed in eliminating the need for abstract mathematical objects, and thereby dispose of the question of how we know such objects, they typically face equally problematic epistemic questions. Yet the authors of these programmes have claimed just the opposite—to wit, that their programmes show promise of being more epistemically tractable than mathematical realism.[1]

I do not know how to prove rigorously that anti-realists inevitably find themselves in this plight, but it does stands to reason. For suppose for a moment that each branch of science has a clear

[1] I examine the epistemic problems confronting realism in the next chapter.

non-mathematical, observational content.[2] Since the chief role mathematics plays in science is that of conceptually organizing and deductively systematizing the non-mathematical, observational content of science, successful anti-realist approaches must deliver the expressive and inferential apparatus that mathematics provides for science (or something at least as strong). But to do this they must introduce some non-mathematical apparatus which goes beyond the observational content of science and is logically strong enough to substitute for mathematics. Now the difficulties with giving a plausible realist epistemology for mathematics arise from the apparent lack of ties between the mathematical apparatus and observation. Thus, because anti-realist substitutes for mathematics must also go beyond observation in ways that are logically similar to the way mathematics does, it is likely that they too will face similar epistemological difficulties.

The foregoing argument is vague because it does not specify the nature of the relative logical strength and similarities between mathematized science and its anti-realist substitutes. This deficiency is due to significant differences in the 'logical' apparatus of standard mathematics and the anti-realist substitutes for it. Standard mathematics uses the extensional logic of the quantifiers 'all' and 'some'. Anti-realists typically trade outright commitment to abstract mathematical objects by extending this logic to include modal operators, or mereology (the logic of parts and wholes), or the logic of stronger quantifiers, such as 'there are finitely many'. As a consequence the anti-realist systems do not fall under the scope of the concepts mathematical logicians have developed for comparing the strength of axiom systems. In the end, the best I will be able to do is to point out, case by case, the epistemological debts of the anti-realist approaches.

1. SYNTHETIC SCIENCE: FIELD

The first approach I will consider is the simplest in its most general conception, but the most difficult to carry to completion: to refute

[2] This supposition could be problematic because mathematical vocabulary can even occur in 'observation sentences' such as 'The computer screen contains a greater number of blank spaces than marked ones' or 'The elevator's downward acceleration is zero'.

the indispensability thesis simply show how to do science without using mathematics. This is the tack Hartry Field took in *Science without Numbers*. He drew his inspiration from Hilbert's *Foundations of Geometry* and the work on measurement theory codified in *Foundations of Measurement* by Krantz, Luce, Suppes, and Tversky. The leading idea of the latter work is that using a certain number system for representing and measuring the data used in a given area of science presupposes that the data exhibit a relational structure that can be embedded within the mathematical structure being applied. For example, on this approach, using the real numbers to measure certain bodily lengths on a ratio scale depends upon these bodies standing in a relation of comparative length (the *longer than* relation) that is a weak order, monotonic under bodily juxtaposition, and so on. Furthermore, where such (empirical) structures are absent, so is the corresponding possibility for measurement. Thus we cannot treat putting soap bubbles together as an operation supporting an additive measure, simply because combining two soap bubbles is unlikely to yield a third one at all, much less one whose size is in any reasonable sense the sum of the first two.

In practice, we make no clear distinction between 'empirical' structures and the mathematical structures in which we embed them. No practical purpose is served by distinguishing between, say, the numerical *greater than* relation holding between numerical values of a length function and the empirical *longer than* relation. But for theoretical or foundational purposes it may be worth attempting to characterize a target structure in terms that do not make use of coordinate systems or numerical scales. Since such characterizations contrast with their numerical counterparts as do synthetic and analytic versions of geometry, the former are commonly referred to as synthetic and intrinsic, the latter as analytic and extrinsic.[3] Euclid's geometry is a synthetic geometry. The early non-Euclidean geometries, which simply used axioms contrary to the parallel postulate,

[3] An extrinsic characterization of a structure refers to a representation of the structure in some other structure. For instance, extrinsic characterizations of spatial structures use coordinate systems. Intrinsic characterizations refer only to elements of the structure or constructions built from them. Analytic characterizations refer to numbers, functions, and sets; synthetic characterizations contain no such references. Field's characterization of Newtonian space-time is both synthetic and intrinsic. On the other hand, a characterization of a structure in terms of tensors could be intrinsic yet analytic.

were also synthetic. Descartes supplied an analytic version of Euclid's geometry, and we now have analytic versions of the early non-Euclidean geometries. There are also synthetic characterizations for certain space-time theories, measurement theory, and utility theory. In each case, we can prove so-called representation theorems that show that the synthetically specified structures can be embedded in their analytic images.

(By the way, such theorems are not necessary for applying mathematics. The structuralist approach expounded by Krantz et al. maintains that the target domain must carry the appropriate structure, to be sure, but it need not be characterized in synthetic or extrinsic terms. Indeed, currently there is no synthetic version of the geometry used in relativity theory. Finally, the approach of Krantz et al. does not undercut the indispensability argument or favour mathematical anti-realism.)

Field hoped to demonstrate that we can expunge mathematics from science by replacing analytic, mathematized scientific theories by synthetic versions that do not refer to mathematical objects. To illustrate how this might be done he formulated Newtonian gravitation theory within synthetic Euclidean geometry.

Although formulating a synthetic version of a branch of science shows how to achieve the conceptual power of (analytic) mathematics, we still might need it in deriving consequences from synthetic scientific statements. Field planned to cover this base by appealing to representation theorems. These theorems enable us to represent a synthetically presented structure (science) in some standard mathematical structure (mathematized science). Once we do this we can use ordinary mathematical reasoning to derive properties of the representing structure, and know that, under the representing translation, these properties transfer to the represented domain in the guise of synthetic descriptions. Thus the mathematics used imputes no synthetic properties to the represented structure, which are not already logical consequences of its synthetic description alone. This means that, contrary to the indispensability thesis, we need not assume that (analytic) mathematics is true in order to use it; for it can be regarded as just a short-cut for generating logical consequences of synthetic scientific statements.[4]

[4] Field's programme encountered a serious technical impediment at just this step. In general, the mathematically derived consequences are logical consequences of the

In assessing Field's and other anti-realist programmes we should ask two key questions: (1) Does the programme succeed in showing that science can be done without appealing to mathematical objects and truths? (2) Is the epistemology for the anti-realist apparatus likely to be any more tractable than the epistemology for a realist's construal of mathematics? When it comes to Field's programme, several considerations indicate that the answer to both questions is negative.

First, the expert opinion is that we have little evidence that Field's approach can even get started with the job of reformulating quantum mechanics or general relativity theory. This is because currently there are no synthetic versions of their underlying mathematical frameworks, and it is not clear that getting them is just a matter of effort.[5]

Second, Field's account is concerned with how so-called pure mathematics might be applied to so-called *mathematical* physics, economics, biology, etc. This is no different from applying one branch of pure mathematics within another—no different, for example, from using number theory or set theory in mathematical logic. But mathematics is also needed to apply synthetic theories to experimental data. Consider, for example, the claims and methods of cardiology or botany, which are about as non-mathematical as they come. These sciences use statistical significance testing to determine whether data are more than coincidental, and statistical estimation theory to determine the values to associate with sets of measurements. Thus cardiologists might use significance testing to see whether the (let us suppose) slightly lower rate of heart attacks among philosophers was just a fluke, and a botanist might use estimation techniques to use the information obtained from a sample to determine the average height of all the trees in a forest. Neither

synthetic theory only if its underlying logic is at least second-order and logical consequence is relativized to standard models. First-order synthetic theories may fail to pick out a sufficiently narrow class of structures to support a representation theorem. Furthermore, both first- and second-order proof methods are subject to Gödel incompleteness results. Thus mathematics will enable us to derive some synthetic consequences of a synthetic theory that cannot be derived using its current proof methods. Thus it is not plausible to see mathematics as a dispensable short-cut. See Shapiro (1983a), and Field (1989), ch. 4.

[5] Balaguer (1996a) is an initial, but still incomplete, attempt at providing a synthetic version of quantum mechanics.

Field nor Krantz et al. address the question of how we are to deal with this kind of applied mathematics.

Experimental physicists use statistics just as much as cardiologists and botanists do. I have used the latter as examples to remind us that even observational sciences require mathematics. Now we don't use statistics to conclude that one counter-example falsifies a universal claim. Rather, we use it to determine whether a set of examples falsifies a statistical claim. One surviving pine tree can falsify the claim that no pine tree survives a pine beetle infestation, but many survivors may fail to falsify the claim that 90 per cent of pine trees do not survive such attacks. This is why statistical inference is an essential part of scientific method. Yet Field's treatment of science says nothing about it.

What is more, it is not clear how Field might get started on meeting this deficiency. The standard account of why it is reasonable to reject a statistical hypothesis on the basis of 'statistically significant' contrary data is that the data are just too improbable given the truth of the hypothesis. Field cannot talk about probabilities, however, because they are numbers; and talking about chances instead will not help, since these are abstract entities too. So he will need to find a synthetic substitute for statistical hypotheses and a different way of accounting for statistical inferences. Again it is hard to see how to proceed. Depending upon how one formulates one's statistics, an experimental distribution is improbable relative to a *sample space*, a *set of events*, a *set of propositions*, a *set of possibilities*, and so on, all of which are mathematical constructions that start with abstract entities.

In the past Field has confronted mathematical challenges that are not directly amenable to a synthetic reformulation by expanding his logical apparatus. It would be natural at this point to add probability or statistical quantifiers to his logic. But it not clear that this alone will do. Suppose that we try to construe '90 per cent of A are B' as a quantification, of the form, say,

$$(90\%\, x)(Ax \rightarrow Bx)$$

parallel to the universal quantification

$$(\forall x)(Ax \rightarrow Bx).$$

Since '$Ax \rightarrow Bx$' is true of everything that is not an A, it could be that '$Ax \rightarrow Bx$' is true of 90 per cent of the things in our universe just because more than 90 per cent of the universe consists of non-*A*s. Thus '$(90\%\ x)(Ax \rightarrow Bx)$' could be true although only 10 per cent of the *A*s are *B*s. To properly analyse statistical claims, Field would need to add non-truth-functional conditionals to his logic—perhaps, even probabilistic conditionals.[6]

Furthermore, since Field could not construe statistical claims as sentences about numerical ratios, it is likely—again judging from moves he has made in the past—that he would take statistical quantifiers or conditionals as primitives. But that would mean that his language would need a different logical operator for every real number! It is hard to see how a nominalist could think of this as a *language*.

Also think of the difficulties in articulating the rules of inference. Given premisses stating that a sample of size k contains m A that are B and $k-m$ that are not B, we need a rule telling us whether to reject (at a certain significance level) the hypothesis that x per cent of A are B. Since the conclusion will vary not only with the particular statistical claim but also with the sample size, mathematics would surely be necessary to give a finite articulation of the rule.

So far I have been discussing Field's plans for eliminating mathematics from *science*. However, Field also needed to eliminate mathematics from *metamathematics*, since rigorous versions of his claims about applying mathematics to synthetic sciences fall squarely within that domain. To purge this mathematics, Field first reformulated his key claims about applying mathematics as claims about the consistency of various theories, and then proposed identifying consistency with logical possibility. The idea here is that instead of saying that some (finite) set of axioms $A_1, A_2, \ldots, A_k$ is consistent, we say that it is logically possible that $A_1\ \&\ A_2\ \&\ \ldots\ \&\ A_k$.[7] On this approach instead of saying that certain mathematical objects, say, the natural numbers, exist, we need only say that it is logically possible that they exist. At least on the face of it, this no more commits one to mathematical objects than saying that unicorns are possible commits one to unicorns.[8]

[6] I owe this point to Adrian Moore.

[7] Field extends this axiom schema by using substitutional quantification. See Field (1984).

[8] Field takes logical possibility as a primitive modal operator in order to avoid committing himself to possible worlds. In order to get this to work out he embeds standard mathematics within a framework of modal operators, modal logic and cer-

Since Field is not the only anti-realist to use primitive modal operators, I will postpone discussing his introduction of modality until the end of this chapter where I can treat it as a general anti-realist ploy.[9] I will argue that it has not resulted in any clear epistemic gain.

Another issue I will address later (Chapter 6) concerns Field's use of space-time theories committed to points and regions. Not only has it been traditional to regard points and regions as mathematical entities, it is also not clear that we have any greater epistemic access to them than we have to numbers or sets. For now let me conclude by noting that we have seen a number of reasons for doubting that Field can complete his technical program whatever its potential epistemic gains may be.

2. SAVING THE MATHEMATICAL FORMALISM WHILE CHANGING ITS INTERPRETATION: CHIHARA AND KITCHER

In parrying the indispensability thesis, Field combined the strategy of eliminating some branches of mathematics from science with showing how to use others (e.g. metamathematics) without presupposing mathematical objects. Charles Chihara and Philip Kitcher take a more conservative approach: by interpreting mathematics in anti-realist terms, they show how we can reaffirm it and mathematized science without committing ourselves to mathematical objects.

2.1. *Chihara's Constructibility Theory*

Since at least Euclid's time mathematicians have found it natural to think and speak of mathematical objects as constructions. Doing so fits Euclidean geometry quite well, where many of the theorems are concerned with constructing figures using a ruler and compass. It is also quite congenial when thinking about proof theory, where many

tain principles relating possibility and consistency. The technical details will not matter here. The technically inclined reader should consult Chihara (1990) where it is argued that Field will still require mathematics in carrying out his possibility proofs.

[9] For further discussion of Field's views see Burgess and Rosen (1997), Chihara (1990), Resnik (1985a), and Resnik (1985b). For expert opinion on synthesizing advanced physics see Malament (1982) and Hellman (1989).

theorems, such as the deduction theorem or the cut elimination theorems, proceed by showing how to convert one formal derivation into another. But mathematicians also speak of the applying a Henkin construction or forming an ultra filter or taking a limit, where, strictly speaking, the methods in question definitely exceed the bounds of constructive mathematics in the technical sense.[10] Thus, although thinking of mathematical objects as constructions comes easily and may actually further mathematical research, it is not a decisive reason for thinking of mathematical objects as things that we literally construct. Not only do mathematicians use the language of construction in applying non-constructive methods, it is also implausible to think of ourselves as actually able to construct most of the numbers, derivations, and other objects treated in constructive mathematics proper. Their magnitude and complexity simply exceeds our computational capacities.

Charles Chihara, whose antipathy to abstract mathematical objects is long-standing,[11] does not believe that we construct mathematical objects. Indeed, since he does not believe in mathematical objects, he can't believe that we construct them. But he does show how to reinterpret mathematics as a theory of constructibility. The first step is to reduce mathematics to the simple theory of types using techniques developed by Frege, Russell and Whitehead, and other logicists. The next step is to replace talk of sets with talk of the open sentences associated with them, while simultaneously construing assertions that an open sentence exists as an assertion that it is constructible. Reinterpreted thus, mathematics no longer commits one to various sets but rather to the constructibility of various open-sentences. (An open-sentence is simply a sentence with one or more free variables, such as '*x* is an even number'.)

Sticking to one language for formulating open-sentences would severely curtail Chihara's reinterpretation, since the number of open-sentences available in a language is countable while the number of sets is uncountable. In fact, Chihara has a more stringent problem to overcome: he recognizes only open-sentence tokens, since untokened open-sentences are abstract entities and no more acceptable to him than sets themselves, and there are only finitely many open-sentence tokens. To get beyond this, Chihara turns to

[10] See Shapiro (1989) for a discussion of dynamic language in mathematics.
[11] Chihara (1973).

open-sentences that *might* be tokened, or as he prefers, that are constructible.

But even this is not enough, since the number of open-sentences that might be tokened using extant languages is countably infinite. So Chihara stipulates that to say that an open-sentence is constructible is something like saying that it is tokened in some possible world (by some being using a language that we might not even be able to grasp or understand). This is supposed to open up resources sufficiently vast and rich to make sense of the constructibility of uncountable infinities. Talk of possible worlds is simply a heuristic, however, since, for Chihara, possible worlds are also unacceptable abstract entities. In the end he takes the locution 'is constructible' as a primitive notion.

Chihara does not run into Field's problem of showing how to purge science of mathematical formalism, because he keeps the formalism while reinterpreting it. In effect, Chihara's response to the indispensability thesis is to concede that asserting sentences that *look* like mathematical principles is indispensable to science, and then to show how to read these sentences so they do not presuppose the existence of mathematical objects. This response can be applied to statistical inference as well—although Chihara does not explicitly deal with this challenge to the project of nominalizing science. Despite this it is not clear that Chihara has shown that his approach can do away with mathematical objects entirely. For he uses mathematics in justifying his system. To see how, consider what he writes in justifying his abstraction axiom: 'To show this, I first define the index of a formula to be the number of occurrences of connectives and/or quantifiers. I then proceed by induction on the index of formulas to show . . . .'[12] Now Chihara might reply that the purpose of this justification is to prove to realists in realist terms that his system suffices for the mathematics needed in science. If so, it would still be important for him to show that his system can be justified in terms that he and other anti-realists can accept. Or he might reply instead that we should reinterpret his mathematical talk in terms of constructibility. But this would call for a further constructibility theory—a metaconstructibility theory—since he is trying to justify his initial constructibility theory. It would then be fair to ask for the

[12] Chihara (1990), 66–7.

justification of this metatheory. Presumably, Chihara would use induction to justify this theory, and we then would press him to eliminate it. This would lead to a meta-metaconstructibility theory, and we would be off on a regress of justifications and reinterpretations. Furthermore, since Chihara accepted the task of justifying his theory in the first place, it would seem that he should do so in terms that are acceptable to those sceptical of constructibility. In that case, Chihara should be obliged to stop the regress at some point and give a neutral, non-mathematical justification of his system.

Let us assume that this can be done, and turn to the question of whether Chihara's approach promises an epistemic gain. We have to speculate on this on our own since Chihara has not produced an epistemology for his theory—at least not as of this writing.

I will start with something Chihara writes in criticizing realism: 'I want to treat the question of the existence of mathematical entities as a scientific question . . . Where is the evidence that supposedly supports belief in these entities? Here, I mean evidence of the sort that would convince the physicist or biologist.'[13] This passage prompts me to wonder about the evidence we might seek for constructibility claims. Take the case of the axiom of choice. Its interpretation in Chihara's system would deny that an open-sentence is constructible where no open-sentence serving as its 'choice function' is constructible. Chihara does not consider the axiom of choice, but it is now part of standard mathematics, and is required for some theorems that are employed throughout science. Thus it should be given a correlate in his system.

Now the considerations mathematicians have historically raised for and against the usual form of the axiom are irrelevant to its constructibility version. One reason is that interpreting constructibility as broadly as he does commits him to disallowing the traditional constructivist objection to the axiom to the effect that specifying some choice functions might require means that we could not grasp. On the other hand, the usual arguments in favour of the axiom to the effect that it is mathematically fruitful to have such an axiom are also irrelevant. Considerations of mathematical fruitfulness, even when translated into terms of constructibility theory, have nothing to do with the sorts of open-sentences that might be tokened.

[13] Chihara (1990),189.

Instead, in gathering evidence bearing upon his version of the axiom of choice, one would think that Chihara would to turn to biology, anthropology, and linguistics in order to determine what sorts of languages are possible and what sort of open-sentences might be tokened.[14]

I am assuming, of course, that Chihara's constructibility theory is concerned with what it is logically possible for language users to do. This is borne out by some of the things he writes in justifying other axioms. Consider his treatment of his correlate of the usual axiom of abstraction. In a simplified form it postulates that for any object *y* and any condition '. . . *x* . . . *y* . . .' formulated in the constructibility theory, an open-sentence is constructible that is satisfied by just the (constructible) things *w* that are such that . . . *w* . . . *y* . . .. Of course, '. . . *x* . . . *y* . . .' is an open-sentence. But it does not verify Chihara's axiom, because it does not *mention* the object *y*. The letter '*y*' occurs in it as a free variable, and in stating the axiom it is used to refer to a specific but (arbitrarily chosen) object. So we need to show that an open-sentence referring to this object is constructible. Here is how Chihara does this:

> [Suppose that in '. . . *x* . . . *y* . . .' the variable '*y*' refers to an object *k*.] Then it is reasonable to maintain that there is some possible world in which the language of this theory is extended to include a name of the object *k*. Then the formula expressing the condition in question can be converted into the required open-sentence by replacing all free occurrences of ['*y*'] by the name of *k*, and surely it is possible to do this.[15]

I am not sure what to make of this. Chihara tells us that he talks of possible worlds only to help us understand constructibility, but here he uses his intuitions concerning them ('it is reasonable to maintain that there is some possible world') to justify his axioms. I do not see how he is entitled to this, and I find his assertion no more convincing than he would that of the realist who said that it is reasonable to suppose that there are numbers having some property.

Furthermore, I find that my intuitions, in so far as I can clarify

[14] Some logicians, mathematicians, and linguists have written of languages with either infinitely long or uncountably many sentences. I presume that, as purely mathematical theories, they are out of bounds for Chihara. Furthermore, he must be wary of idealizations of ourselves, since he complains that Philip Kitcher's ideal constructor 'appears more godlike than human' (Chihara (1990), 243).

[15] Chihara (1990), 66.

them, conflict with Chihara's. Guiding myself by his earlier statement that we are to imagine a possible world 'in which some *people*, who have an appropriate language, do something that can be described as the production of the token',[16] I try to imagine what is humanly possible. And it seems to me that there may be physical objects that it is not humanly possible to name, simply because it is not humanly possible to identify them with sufficient precision. They might be too small, too fast, or too fleeting. (Chihara seems to have no qualms about physics; so he should not object to examples from quantum field theory, which posits processes of such short duration that they are undetectable.) On the other hand, if Chihara simply means that for any object k it is logically possible that some being tokens a name for it, then his intuition seems more plausible.

In the end, then, Chihara's epistemology amounts to the epistemology of logical possibility. Now once we give up mathematics as a tool for determining logical possibilities, we seem to be left with three resources: modal logic (to tell us what is possible if . . .), inferences from what we know to be true to the possibility of claims of the same logical form, and logical intuitions. The first two are probably too weak to provide all the knowledge of logical possibility Chihara's project requires, since it is likely that they cannot show that indefinitely complex expressions are possible. I put little stock in the last, which is connected to the second through including intuitions concerning logical forms, since often even the intuitions of professional logicians conflict. Finally, I doubt that there are modal facts to be known—even when the modality is that of logical possibility. But more on this in Chapter 8.

2.2. *Kitcher's Idealizations*

In *The Nature of Mathematical Knowledge* Philip Kitcher offers an account on which mathematics is an 'idealizing theory' like the physical theories of ideal gases, mass points, and frictionless surfaces. These theories idealize various features of the observable world. Kitcher's mathematics idealizes both the world around us and our ability to perform the most basic operations used in practical mathematics: segregating and matching (used in numbering), measuring, cutting, moving, and assembling (used in geometrical construc-

[16] Chihara (1990), 40, my emphasis.

tions), and collecting (used in forming sets). On this view, number theory, for instance, is a theory of a perfect counter, who lives forever without suffering lapses of memory or attention in a stable universe of infinitely many objects.

Physicists know from the outset that neither ideal gases, mass points, nor frictionless surfaces exist, but by investigating their properties they can derive conclusions that may usefully describe the approximate behaviour of real gases, bodies, and surfaces. Despite this, the use of ideal theories seems problematic. If their generalizations are formulated in standard extensional logics then they are vacuously true, since they speak of non-existent entities and thus have no counter-examples. By the same token, their contraries are also vacuously true: both 'all balls rolling down a frictionless plane reach the bottom' and 'no balls rolling down a frictionless plane reach the bottom' are true.

Thus one might object that Kitcher's approach will yield too many truths. We can set this worry aside, however, because we are not interested in what is true of all ideal entities (every generalization is!) but rather *what is true according to a given ideal theory*. This is a matter of what follows logically from the basic assumptions of the theory. Similarly, we do not want to know what is true of all ideal counters, but rather what is true according to the theory of them. Thus, if someone asks why in an ideal gas the pressure increases, rather than decreases, as its container collapses, we can respond that the ideal theory implies this. And if someone asks why the molecules of an ideal gas are unextended, we can respond that this is part of the definition of an ideal gas. Similarly, Kitcher can point out that ideal counting is invariant under re-ordering the objects counted because the principles that define ideal counting imply that it is.

In science we formulate our ideal theories against a mathematical background and use both mathematics and logic to determine what is true according to them. But Kitcher wants to treat mathematical theories themselves as ideal. So he must use only logic as his background inferential apparatus. This is no impediment to developing his ideal theories, since something is true in a given mathematical theory if and only if it is a logical consequence of its axioms.

Can Kitcher continue to stick with purely logical inferences when it comes to applied mathematics? This is a bit complicated, because it is not clear whether his idealizations are normative, defining correct counting, for example, or descriptive, describing how we actually count when we are not distracted, and so on. To see what I mean, suppose that we construe Turing machine theory as an ideal theory of actual computing machines. Assume that we use a Turing machine *T* to design a type of computer *C*. Let us further suppose that we can prove that *T* never prints the symbol '1', and that we build a computer which is supposedly of type *C* and it prints a '1'. Then we can conclude using just logical inferences that the computer in question failed to conform to its specifications. Here we use Turing machine theory normatively. But presumably we want to be able to use the behaviour of *T* to predict how machines of type *C* are likely to behave. If so, then we will need to know how reliable these machines are, how, where, and when they are likely to fail, and so on, and this will require us to use statistical techniques. So we will need mathematics to apply Kitcher's anti-realist version of mathematics to empirical data—just as Field did.

This is not all. Ideal theories must be consistent, for according to an inconsistent ideal theory all assertions about its ideal objects are true. Inconsistent idealizations bring back the problem of vacuous truth with a vengeance. But we need mathematics to state and prove that a theory is consistent, since idealizing the mathematics of consistency proofs would rob them of whatever conviction they might carry. Now we have already seen a way out for Kitcher, though he did not use it (or address the issue of consistency): he can trade talk of consistency for talk of logical possibility.

But before turning to assessing this trade I want to discuss briefly Mark Balaguer's recent argument that we can account for the applicability of mathematics in science without having to acknowledge mathematical objects or truths, or reformulate science or reinterpret mathematical claims.[17] Balaguer does not directly address the indispensability thesis or either indispensability argument, because he focuses on *accounting* for the applicability of mathematics. We have been concerned instead with the problem of carrying out scientific inference without presupposing the truth of mathematics. Despite this, we can discern a position on our problem in his remark that fictionalists can maintain that mathematics serves as a heuristic device for understanding physical phenomena just as a

[17] See Balaguer (1996b).

novel or a play can be used to help us understand historical phenomena. The idea here is that in using a mathematically formulated scientific theory in modelling some physical phenomena, we need not take the theory to be true. Contrary to the indispensability thesis, we need only assume that the physical phenomena are *like* the mathematics-based theory in order to draw conclusions from it.

The idea I am attributing to Balaguer is like Kitcher's without the detour through the ideal constructor. And it is open to similar objections. First comes the question of how we are to compare scientific models with the world. If we use statistics, then we can avoid assuming that our scientific theories are true and still draw certain conclusions from them. But such statistical modelling presupposes the truth of the mathematics used in statistics. Now Balaguer might point out that in reasoning from features of a model to features of the world we can use analogical reasoning in place of statistics. To be sure, this would dispense with the mathematics of statistics, but it would also dispense with those parts of science where statistics is *de rigueur*. Thus this proposal would refute the dispensability thesis only by restricting the domain of science.

A second problem concerns the consistency of our mathematics-based scientific theories. As we saw, even if we need not suppose that these theories are true, we want them to be consistent. Presumably, Balaguer would substitute logical possibility for consistency and fall under the same rubric as Field, Kitcher, and our next protagonist, Geoffrey Hellman.

3. AN INTERMEDIATE APPROACH: HELLMAN'S MODAL-STRUCTURALISM.

Geoffrey Hellman develops a view of mathematics which combines some of Field's ideas on applied mathematics with an anti-realist interpretation of pure mathematics composed of ideas found in the works of Dedekind and Putnam.[18] Hellman calls his position 'modal-structuralism', because he uses modal operators to translate standard mathematical statements into a 'structuralist' language, in which he substitutes assertions of logical possibility for mathematical existence claims. Thus he does not assert, for example, that

[18] Hellman (1989).

infinite progressions (omega sequences) exist, but only that it is logically possible that they exist. Furthermore, to avoid introducing abstract objects in his metatheory, he eschews the possible world semantics and takes modal operators as primitive—in this way mirroring Chihara and Field. For pure mathematics, his background logic is second-order S5, in which all iterations of modal operators collapse to either a single possibility or a single necessity operator and quantifiers binding predicate and function variables are available. Moreover, at least in the case of number theory and analysis, he suggests that we can think of the relevant possibilities as realizable by systems of concrete, material objects (for example, an omega sequence of inscriptions). This in turn suggests that Hellman's structuralism also has an Aristotelian twist to it; the only actual structures are those that are concretely realized.

3.1. *Number Theory and Analysis According to Modal-Structuralism*

Hellman's treatment of pure number theory and analysis shows his project at its best. For here he is able to honour his realism by producing bivalent, modal-structuralist translations of ordinary mathematical statements, and here his epistemic, metaphysical, and modal commitments are simplest.[19] For brevity and simplicity I will restrict my exposition to his treatment of number theory.

To see how this works, let *S* be any closed sentence of Peano number theory (including second-order sentences quantifying over number-theoretic functions, classes, and relations). In this system addition, multiplication, and the other familiar number-theoretic functions can be defined in terms of the successor function, so we can assume that the only non-logical symbol appearing in *S* is a symbol for successor. Let *PA2* be the conjunction of the second-order Peano axioms. As Dedekind first proved, these axioms are *categorical*, that is, their models are isomorphic to each other. As a result the same models are also *elementary equivalent*, that is, *S* is true in all the models of *PA2* or in none. Thus, *S* is true (in the objects-platonist's standard model)[20] if and only if '*PA2* $\rightarrow$ *S*' is a

[19] I will not discuss his treatment of set theory here, where honouring these commitments becomes much more difficult, but see Resnik (1992b), from which much of the present discussion of Hellman's view is taken.

[20] Hellman uses the term 'objects-platonist' to contrast the standard realist (pla-

second-order logical truth. Now restrict all the first-order quantifiers in the last sentence to the new free, second-order predicate variable 'X',[21] and replace the successor symbol everywhere by the new free, second-order function variable 'f'. Finally, universally quantify these two free, second-order variables. This yields a sentence of pure second-order logic, which I will represent as

$$(\forall X)(\forall f)[PA2 \rightarrow S] \mid X;f.$$

(The notation '$\mid X;f$' is short-hand for the result of substituting 'X' and 'f' in the appropriate places in the sentence preceding it. The entire displayed formula could be glossed as 'in any domain X supporting a function f, if the Peano axioms hold for the pair $\langle X;f \rangle$, S also holds for $\langle X;f \rangle$'.) This sentence is true in second-order logic just in case the original sentence S is a truth of (second-order) number theory. If, with Hellman, we equate second-order logical truth with logical necessity, then we can see that S is true just in case

$$(S_{ms})\ \text{Nec}(\forall X)(\forall f)[PA2 \rightarrow S] \mid X;f.$$

This is Hellman's modal-structuralist translation for S.

If the antecedent of S_{ms} is impossible, then it will be vacuously true. Thus Hellman faces a problem similar to the vacuity problem Kitcher faces. Neither can simply assume that their antecedents have models, since doing so presupposes mathematical objects. But they can ensure the non-vacuity of their translations by assuming that such models are possible. Hellman does this by postulating

$$\text{Pos}(\exists X)(\exists f)PA2 \mid X;f.$$[22]

The technique used to develop a modal-structuralist version of number theory also applies to second-order real analysis, whose individual variables range over real numbers, and to other mathematical theories, such as Euclidean geometry, that have finite, second-order axiomatizations yielding internal categoricity and elementary equivalence theorems. Of course, to guarantee non-vacuity we must

tonist) view of mathematical objects with the view of non-modal structuralists such as Stewart Shapiro and myself.

[21] For example, '$(\forall n)(\ldots n \ldots)$' becomes '$(\forall n)(X(n) \rightarrow (\ldots n \ldots))$'.

[22] It turns out that Hellman is able to derive this 'modal existence' postulate from another more 'constructive' axiom, which he dubs 'potential infinity'. In effect, it asserts that it is logically possible for there to be an infinite progression of concrete marks.

add further modal existence postulates, for example, for real analysis we might postulate the possibility of a continuously ordered field.

Hellman would like the foundation given so far to count as nominalist. The sticking point concerns the second-order variables, which are usually taken to range over classes, properties, or Fregean concepts. Hellman points out that if we count as concrete certain geometric models of number theory and analysis, then we can satisfy nominalist demands by construing the second-order variables as ranging over mereological sums of *concreta*. Thus provided we acknowledge the concreteness of geometric objects, we can represent 'an enormous amount of mathematics . . . in terms of concrete atomic structures with concrete Cartesian spaces' without committing ourselves to actual abstract entities.[23]

3.2. *The Modal-Structuralist Account of Applied Mathematics*

In treating pure mathematics modal-structuralists posit structures as mere possibilities without asking what their realization might require. Applying mathematics in science, technology, and practical life forces one to look at the world as it is. This makes modal-structuralists connect the possibilities introduced at the level of pure mathematics with the actual world—at the unfortunate price of introducing additional modal operators.

To see how this arises consider the simple applied mathematical sentence, 'The number of stars is finite.' As a first pass, Hellman suggests that we translate this in modal-structuralist terms by

$$\text{(Star)}\quad \text{Nec}(\forall X)(\forall f)[PA2 \rightarrow (\exists n)(X(n)\ \&\ (\exists g)(g \text{ is } 1\text{–}1\ \& \\ (\forall x)(\text{Star}(x) \rightarrow g(x) < n)))] \mid X;f,$$

or, as he would read it informally,

> If there were a standard model of second-order number theory, then there would be a one–one mapping from the stars into some initial segment of the model's natural number sequence.

[23] Hellman (1989), 51. Hellman suggests that these models could use space-time points to play the role of real numbers and some extra individuals to play the role of ordered *n*-tuples. In a more recent paper he proposes interpreting second-order quantifiers plurally in the manner of George Boolos in order to avoid commitments to second-order entities. See Hellman (1996) and Boolos (1984).

Hellman emphasizes that counterfactual claims of this type must be construed as concerned with the actual world. If we allowed entities in other possible worlds to serve as numbers, then the counting function referred to in (Star) would be a trans-world mapping, that is, an intension, and fall outside the scope of the modal-structuralist mathematical framework.

Thus for this translation to be non-vacuous it must be possible for the actual world to be augmented with entities modelling *PA2*. This raises an additional problem; for we must also be assured that the possibility in question is one which leaves the number of stars exactly as they actually are. For example, it will not do to use an infinite progression of *stars* as a model, since that would entail the falsity of the modal-structural translation of our example regardless of the actual number of stars. To deal with this, Hellman inserts a clause into the antecedents of his translations excluding such 'interfering' models, and lays down modal existence postulates that posit the possibility of models that would not interfere with the way the world actually is were they to exist. Thus we might rewrite (Star) as

(Star′) Nec$(\forall X)(\forall f)[PA2$ & $X;f$ does not interfere with the way things actually are $\rightarrow (\exists n)(X(n)$ & $(\exists g)(g$ is 1–1 & $(\forall x)(@\text{Star}(x) \rightarrow g(x) < n)))] \mid X;f$,

glossed as

> If there were a standard model of second-order number theory that did not interfere with the way things actually are, then there would be a one–one mapping from the stars there actually are into some initial segment of the model's natural number sequence.

But notice that (Star′) contains a new modal operator, '@' (read 'it is actually the case that'), which, speaking intuitively, takes us from any possible world back to the actual world. Without it we cannot guarantee that we are referring to the stars in our world. Furthermore, if modal-structuralists want to extend their treatment to sentences dealing with 'other worlds', such as

> Even if the number of stars is not actually finite, it might have been finite,

then they will need an operator (some call it 'the backspace') which, again speaking intuitively, takes one back to the last world designated.

(Surely, once one takes the modal leap, one will want to allow claims of this type too.)

Neither of these operators is definable in terms of the usual possibility and necessity operators.[24] Thus modal-structuralism must pay a steep price in accounting for applied mathematics: namely, invoking a modal operator which cannot be equated with any of the purely logical modalities. Actuality is a metaphysical concept *par excellence*.

Specifying the required non-interference clauses raises the price further. We might take 'does not interfere with the way things actually are' as a new primitive. Or we might try quantifying over ways things are and use the actuality operator. Neither choice would appeal to Hellman. He explicitly rejects the first as too imprecise and subject to the sorts of philosophical objections Nelson Goodman has lodged against talk of the way the world is.[25] He does not consider the second, but clearly he cannot reject quantifying over possible worlds and then quantify over ways things might be or even over properties or relations taken as intensions.

Because of this, Hellman moves onto to explore ways of formulating non-interference conditions in more precise and metaphysically neutral terms. His general strategy is to introduce non-mathematical predicates R and use them to formulate clauses of the form

> $x, y, z, \ldots, w$ stand in relation R if and only if $x, y, z, \ldots, w$ actually stand in R, for all actual objects $x, y, z, \ldots, w$.

Given an appropriate set of predicates, we could use these clauses to state in non-mathematical terms that in a given counterfactual situation all the actual objects have exactly the (presumably relevant) properties and relationships to each other that they actually have.

Hellman notes that, like Hartry Field, he needs to fix a structure via a non-mathematical description of the physical facts, but unlike Field he is not obliged to reformulate applied mathematical theories as 'attractive' synthetic ones. Thus, despite the difficulties with Field's programme, he remains cautiously optimistic about his own proposals for finding synthetic predicates.

[24] For formal details see Hodes (1984). For a valuable discussion of attempts to modalize away commitments to mathematical entities see Burgess (1994). His paper increased my appreciation of the role of the actuality operator in Hellman's translations of applied mathematical claims.

[25] See Hellman (1989), 128.

Whether or not Hellman's suggestion for dealing with this problem eventually works out, his programme, along with Field's and Kitcher's, appears to break down when it comes to statistical inference. Consider the following hypothetical case. Some chemists synthesize a new compound S_n that they believe to be slightly different from an old compound S_o. Using mathematically expressed chemical hypotheses C and mathematical theorems T, they calculate that the melting points of the two substances should be identical. We can represent this by the conditional

$$C \,\&\, T \rightarrow m(S_o) = m(S_n).$$

Then they decide to test C. Taking T for granted, they conclude

$$m(S_o) = m(S_n),$$

which allows them to infer that the arithmetical means of distributions of measurements of melting-points of samples of the two substance should be the same. We could write this as

$$\text{dist}_{am(}S_o) = \text{dist}_{am}(S_n).$$

Then they melt samples of the new substances and record their melting points, and compare the mean value of these numbers with the value to be expected given their hypothesis. Let us suppose that the difference between these is 'statistically significant' at a level they consider appropriate, and, as a result, they take the experiment to refute their hypothesis.

Now observe how Hellman's account would run. The initial calculations establish something like this subjunctive conditional:

> If there were a non-interfering model of second-order real analysis and C held, then $m(S_o) = m(S_n)$ would hold.

Here *non-interfering* means not interfering with the chemical processes relevant to hypothesis C. Of course, this would be spelled out by means of appropriate non-interference clauses. The next step takes us from the melting-points themselves to distributions of measurements of melting-points, and gives rise to something like this:

> If there were a non-interfering model of second-order real analysis, then $\text{dist}_{am}(S_o) = \text{dist}_{am}(S_n)$ would hold,

and thence to this:

> If there were a non-interfering model of second-order real analysis, then samples with means not between *x* and *y* would be significant at the *z* level.

But now two problems arise. The first is finding a non-mathematical substitute for talk about means of distributions. One way would be to introduce predicates describing the tables in which the measurements are recorded. Then instead of talking about patterns (distributions) of numbers we could talk about the patterns exhibited by written tables of numerals.

But the second problem is more serious. The original inference continued with:

> The sample mean is not between *x* and *y*;
> So it is significant at the *z* level;
> So the hypothesis is disconfirmed.

The first two steps could translate as:

> If there were a non-interfering model of second-order real analysis, then the sample mean would be not between *x* and *y*.
>
> If there were a non-interfering model of second-order real analysis, then the sample mean would be significant at the *z* level.

Since the non-interfering clauses would connect conditions on distributions with conditions on tables, we could infer that the data table for the new substance differs 'significantly' (in some non-mathematical sense) from that for the old substance. But now we come to the second problem, which is to explain why the original hypothesis has been disconfirmed.

The standard explanation is fairly straightforward. Given the chemical hypothesis, the distribution of measurements the experiment yielded is just too improbable; so we reject the hypothesis. But modal-structuralists run into the same problem that Field does: they cannot talk about probabilities because they are numbers. And talking about chances instead will not help, since these are abstract entities too. So the modal-structuralists will need to translate the inference by which we reject the chemical hypothesis into modal-structural terms. But what non-mathematical predicates can they use? Depending upon how one formulates one's statistics, the experimental distribution is improbable relative to a *sample space*, a *set of events*, a *set of propositions*, a *set of possibilities*, and so on, all of which are mathematical constructions that start with abstract entities. So I do not see how modal-structuralists can even get started. Until they show us how to overcome this difficulty their attempt to refute the indispensability thesis will remain incomplete.

4. WHAT HAS INTRODUCING MODALITIES GAINED?

Thus far I have argued that neither Field, nor Hellman, nor Kitcher have succeeded in showing that an anti-realist substitute or construal of mathematics can completely replace standard mathematics, because it is unclear that they can deal with statistical inference. Now one might be able to bypass statistical inference by using analogical reasoning. I doubt that Field, or Hellman, or Kitcher would find this a plausible solution.[26] In any case, for the remainder of this section I will simply assume that they have produced some sort of solution, and ask whether we have much reason to think that a modal approach to mathematics represents an ontic and epistemic gain over standard mathematics.

Recall that Chihara, Field, and Hellman are realists about logical possibility. Indeed, they must be if they are going to invoke modal premisses in place of mathematical ones in applications of mathematics.[27] The main difference between them and ordinary mathematical realists comes to this: where the former believe that mathematical objects of a given kind exist, the latter only believe that it is logically possible that these objects (Field) or objects with their structure (Chihara, Hellman) exist.

[26] Chihara seems to require standard mathematics in giving a metatheoretic justification of his system.

[27] Recall that these philosophers do not question the validity of the indispensability arguments but only the indispensability premiss, which they refute by finding substitutes for mathematics. Thus the analysis of scientific reasoning supporting the indispensability premiss can be used to show that they are committed to the truth of their substitutes.

Let me add some qualifications. Field's use of modal premisses will be restricted to metamathematics. Furthermore, I have not included Kitcher with Chihara, Field, and Hellman, because he does not explicitly invoke modalities. Earlier I did so on his behalf in order to provide him with a response to the vacuity objection. This extension of his position would be committed to the truth of modal premisses.

Let us also note that both Field and Hellman are committed to recognizing that abstract entities are logically possible. Field explicitly commits himself to the possibility of sets satisfying the axioms of NBG set theory.[28] Hellman does not go so far. However, after a valuable examination of the mathematics we might need for physics he does conclude that 'it does seem misguided to seek any a priori limit' on the mathematics science uses, including even the use of large cardinal axioms.[29] Furthermore, when it comes to transfinite set theory, he admits that 'we can hardly call [the associated modal-structuralist framework] "nominalistic"'. But then, as if to anticipate my claim, he adds that it is unnecessary to make 'any assumption as to "the nature" of the objects' that might realize these structures.[30] True, no such assumption need appear in the modal-structuralist formalism; but surely the question of the nature of the objects that might be involved is relevant in assessing the philosophical value of the project. Plainly Hellman cannot argue at this point that the objects could be concrete.[31]

[28] See Field (1989), 109.

[29] Hellman (1989), 124.

[30] Hellman (1989), 117.

[31] It is also not clear that Hellman can avoid committing himself to the possibility of abstract objects even when dealing with arithmetic and analysis. In an Aristotelian spirit Hellman often writes as if to add a non-interfering omega sequence or a continuum to the world is to add an arrangement of concrete things to it. Can we do this without changing the behaviour of the things already there? That depends upon what we mean by the actual world—on whether we count just its population and their properties and relations taken in extension, or its laws as well. Adding a new planet to our solar system will not change the behaviour of the other planets so long as we change the physics of the new world. There is no place for the laws of physics in Hellman's sketches of non-interference conditions. This suggests that he may be willing to take the actual-world-*sine*-physics approach to non-interference. Yet we do not want to give up all the physics, since the application in question may be within a theory that presupposes physics. Thus we may need to assume that the physics of the relevant actual objects stays the same, and that the new objects do not affect the behaviour of the old ones. This is a possibility that Hellman admits: 'we are free to entertain the possibility of additional objects—even physical objects . . . [that] could be conceived as occupying a certain region of space-time but as not subject to certain dynamical laws' (Hellman (1989), 97). But once we free our world of its physics to the extent Hellman contemplates, why should we call the new entities physical or concrete? Of course, since he is already in the business of stipulating, Hellman might respond that he stipulated that they are concrete and that's that. But how can he be confident that he knows what he is stipulating?

Philosophers of mathematics frequently assume that the distinction between abstract and concrete (mathematical and physical) objects is clear and unproblematic. Yet Hellman's examples, and Hartry Field's use of space-time points in his nominalization of physics, show that without a clearer grasp of the distinction than we

Thus the distinction between saying that abstract objects exist and saying that it is logically possible that they exist is crucial to the contention that Field's and Hellman's positions represent ontic and epistemic progress. Intuitively, it seems to be the difference between saying that something might be (or that it is consistently describable) and saying that it exists, and this distinction is reasonably clear when it comes to possible *concreta* such as unicorns. But when it comes to sets or structures like the iterative hierarchy, which cannot be concretely realized, the difference threatens to be merely verbal.

This is arguably the case for Platonic (as opposed to Aristotelian) properties, relations, and other universals, which can exist uninstantiated. Now it might be that some universals cannot be instantiated if others are, just as the existence of arbitrary subsets of every set precludes a universal set. (Admitting one would generate Russell's Paradox.) So when we talk of the possibility of instantiating a universal we should understand it as relative to a background system of universals. But given this understanding, I can see no more reasonable condition to demand for a universal to exist than that it is possible for it to be instantiated.

Some philosophers have argued that mathematical objects are universals. For these philosophers there would be no question that the possibility of mathematical objects suffices for their existence. Our anti-realists have not taken mathematical objects to be universals. But they are clearly committed to the possibility of some non-concrete entities exhibiting various mathematical structures, for example, entities exhibiting an iterative hierarchy. Presumably these objects have no positive properties beyond those they have in virtue of their relations to one another. They are—to use a term that will come to the fore in Part Three—simply *positions* in a structure. Granted this, it is again hard to see how the possibility of such objects standing in such relations fails to suffice for their existence.

Let turn now to the epistemology of modal mathematics. We might expect that some advantage is to be gained by replacing the question of how we know that entities exhibiting mathematical structures exist by the question of how we know that such objects are logically possible. However, both Field and Hellman would be

now have we cannot make a definitive assessment of what these projects accomplish. In Chapter 6 I give reasons for doubting that there is a clear distinction between mathematical and physical objects.

among the first to admit that they have little idea of how we know that objects having the structure of an iterative hierarchy of sets are logically possible. But they offer us the hope that the epistemology of logical possibility will prove more tractable than the epistemology of mathematical objects.

I do not share their optimism. First, I am uncertain of what we are supposed to know when we know that it is logically possible that there be so and so. Field suggests that we could define logical possibility in terms of a primitive notion of logical implication and convey the meaning of the latter 'by specifying the procedural rules involved in inferring with it'.[32] This just passes the buck to the procedural rules. How are we to know that they correctly characterize logical implication? Remember Field and Hellman are committed to facts of the matter here.[33] They cannot fall back upon some conventionalist ploy whereby we (partially) fix implication and logical possibility by laying down rules of inference. And what do we know when we know that one thing implies another or that something is logically true? What in addition to knowing that, for example, everything is self-identical do we know in knowing that it is logically true that everything is self-identical? I fear that it may have to be the extra-worldly sort of thing that anti-realists find so objectionable about abstract entities.

Even if the modalists manage to clarify what it is that we know in knowing logical facts, I suspect that coming to know that infinite structures are logically possible will involve the same sort of processes that mathematicians could use in coming to know that they are actual. In short, if we had a satisfactory epistemology for the possibility of mathematical objects, we would already have one for mathematical objects themselves.

Let me elaborate a bit on this. Granted, in most non-mathematical cases it is easier to show that something is possible (or consistent) than to show that it is actual (or true). Undoubtedly this has led modalists to suppose that it is easier to explain how we could know that mathematical entities are possible than it is to explain how we could know that they exist. Let us take Hellman's simplest case, the possibility of an infinite progression. Given our experience with set theory and mathematical logic during this century, we can be confident that we cannot deduce this possibility from our knowledge of

[32] Field (1989), 32.

[33] See n. 27 above.

the finite. Historically, we probably arrived at our belief in the possibility of concrete infinities through untutored physical intuitions. But once we set aside our mathematical knowledge, I do not see how we could count this belief as justified. Observation, biology, and engineering do not tell us that natural, biological, or mechanical sequences can always be prolonged by one more step. Rather they tell us that prolonging a relative short sequence may be quite different from prolonging a very long one, for eventually the process of generating additional steps will exhaust the material needed or the mechanism involved. If we remove these limitations on the grounds that they are merely biological, practical, or technical, then we might be tempted to conclude that it is physically possible to prolong certain natural sequences indefinitely. It is not clear how we could justify that conclusion, however. For to suppose that we could continue such sequences by increasing velocities indefinitely, or by dividing matter into ever smaller parts, or by using ever more matter, energy, space, or time is to presuppose the possibility (perhaps the existence) of potential physical infinities.

Fans of modalism will be quick to reply that the relevant possibility is not physical but logical. Yes, but if our physical experience will not help, how do we know that infinite progressions are logically possible? Because we see no contradiction in the supposition that they exist, I presume. And how do we know this? Well, we can rest with logical intuitions and our deductive experience or we can turn to mathematical models. Historically we have taken the latter course, and have appealed to mathematical objects to clarify intuitive notions of possibility. Thus possible paths and shapes have given way to curves in space, possible sizes, weights, and temperatures to abstract magnitudes, and universal physical possibilities to distributions of matter in space-time. Even in logic we have turned from untutored notions of possibility to mathematical notions, replacing the idea of logical possibility with that of implying no contradiction, and explicating that in turn in terms of proof-theoretically defined deductions or set-theoretically defined models. And why have we done this? Because moving to the more abstract realm of mathematics allows us to put our ideas in their simplest and most uncluttered forms, thereby giving us the best chance of determining their consequences. It is easier to determine whether the idea of the natural number sequence *qua* bare structure of an infinite progression harbours a contradiction than it is to determine whether the idea of some physical infinity does, simply because we need not worry about the effect of extra physical baggage.

Extrapolating from the finite may very well suggest that infinite structures are possible (or exist), but it will not provide sufficient evidence for this on its own. We simply have to posit that they are logically possible (or exist), and test our postulates through their coherence with our prior beliefs and their fruitfulness. Under these circumstances it makes more sense to posit the possibility of objects to which we attribute no ordinary physical properties. (In the case of geometrical objects, we have no choice. An extensionless point is no bit of matter.)

So far I have been arguing that our knowledge that infinite physical structures are possible is likely to depend unavoidably upon our knowledge that infinite mathematical structures are possible. This should not alarm modalists much. But remember Field and Hellman must do more than posit the possibility of mathematical objects; they must also postulate principles that allow them to derive the same non-mathematical consequences we would ordinarily derive using standard mathematics. (Hellman requires non-interference principles and synthetic determination results; Field needs principles embedding his modal primitives within metamathematics.) Since both the standard and modal systems aim to have the same set of non-mathematical consequences, they should imply the same set of observational consequences. In short, they will be empirically equivalent. And since it is likely that both will depend upon the same sort of hypothetico-deductive epistemology, in the end the evidence for standard mathematical systems and the modal systems is likely to be the same. But if that is so, then the modal system, which has a more complicated apparatus than the standard one, cannot have the epistemic advantage. Simplicity is enough to recommend the standard system.[34]

[34] I wrote this paragraph before reading a draft of Burgess and Rosen (1997). However, they make related criticisms of the scientific merits of nominalist systems. The reader interested in an excellent and thorough discussion of both nominalist responses to the indispensability thesis and the reasons motivating them should consult their book.

5. CONCLUSION

Those anti-realists who have not shown that their systems can do the work of standard mathematics have not completed the first step towards undermining the indispensability thesis. Furthermore, although our anti-realists object to mathematical objects as epistemically inaccessible, we can raise similar worries concerning the logical possibilities they require. So even if they could show that their systems are genuine technical alternatives to standard mathematics, they would still have to show that theirs are less objectionable. Finally, the most promising epistemology for possibility—a postulational epistemology—should work as well for mathematical objects. (I aim to establish this over the remainder of this book.)

I have another reason for being sceptical about the modal approach. I am an anti-realist about logical possibility and necessity. Obviously, I need to defend and explain this view, especially since logical deduction has always played a central role in developing mathematical theories, and I will do so in Chapter 8 where I discuss the nature of mathematical evidence.

5

Doubts about Realism

One might well wonder why so many talented philosophers have devoted so much effort to undercutting the indispensability thesis. It is because when we try to assimilate mathematical objects into our everyday conceptual scheme the shortcomings of mathematical realism irritatingly emerge. If we ask, for example, where we can find mathematical objects or how we might perceive them, official mathematics remains mute. Whether or not mathematical objects have locations, smells or colours is totally irrelevant to mathematical practice. Speaking unofficially, mathematicians would likely tell us that it makes no more sense to ask where a number is than to ask what a song smells like, and that our quest is in vain, since mathematics has already told us all there is to say about mathematical objects.

1. HOW CAN WE KNOW MATHEMATICAL OBJECTS?

Mathematics remains silent about the metaphysical nature of its objects, but this does not stop philosophers, for they seek a general theory of the Universe, and they will gladly speculate when scientists hesitate. Mathematics tells them of infinities upon infinities of mathematical objects and of perfectly straight lines and extensionless points. Neither mind nor matter embodies things as numerous or as perfect. As a result, since antiquity many philosophers have concluded that mathematical objects must be, if they are anything, causally inert entities existing outside space and time. I concur with these philosophers. But this conclusion simply generates another question: how could we have ever come to know such strange things? Plato theorized that our souls dwelt in the realm of *abstracta* prior to our bodily births, and what mathematical knowledge we have, we have actually recalled from that time.[1] Frege spoke of

[1] Plato applied this to genuine knowledge across the board.

mathematical objects as being given directly to reason and 'utterly transparent to it',[2] while Gödel wrote that 'despite their remoteness from sense experience, we do have something like a perception also of the objects of set theory, as is seen from the fact that the axioms force themselves upon us as being true'.[3]

None of these pronouncements have been very helpful, since they have remained sketches at best, closer to reaffirmations of the abstractness and independence of mathematical objects than to explanations of our knowledge of them. Frege hoped to show that mathematics is just an elaboration of logic, but his logic had its own abstract objects, and he explicitly declined to say how we come to know the truths of logic.[4] The logical positivists tried to fill in the gap in Frege's epistemology by declaring logic and mathematics to be true by convention. But, on the logical positivists' understanding of truth by convention, mathematical truth and existence are not independent of us and our mathematical activity. On their approach, in stipulating that, say, the axiom of choice is true, we are *making* it true instead of acknowledging an independently obtaining fact. And if we decide that mathematics will do better by denying this axiom then, according to the positivists, we are free to change our conventions and *unmake* its truth.

The failure of so many realist epistemologies to advance beyond brief hints about a priori insights into the mathematical realm has rightly prompted people to wonder whether an adequate realist epistemology is even possible. In a very influential paper, 'Mathematical Truth', Paul Benacerraf raised serious doubts about its prospects. His argument went roughly along these lines:[5]

(1) Some mathematical statements must be known directly, i.e. without being inferred from other statements;
(2) Any plausible account of direct knowledge must appropriately connect the grounds the knower has for knowing a statement with the conditions under which that statement is true;
(3) On the realist interpretation of mathematical statements this means appropriately connecting our grounds for knowing, e.g. that $3 + 2 = 5$ with $3 + 2$ being 5;

[2] Frege (1959), sect. 105.
[3] Gödel (1963), 483–4.
[4] See Resnik (1980), 175 for a supporting quotation and further discussion.
[5] See Benacerraf (1973). This reconstruction owes something to Maddy's discussion of Benacerraf's argument in Maddy (1990).

(4) But our grounds for knowing that 3 + 2 = 5 (or whatever directly known mathematical statements there are) consist in our having certain experiences;
(5) Mathematical objects are causally inert;
(6) So they play no role at all in bringing about those experiences;
(7) Thus our grounds for directly knowing mathematical statements cannot be connected with the conditions under which they are true.

In short, mathematical realism seems to imply that if there are mathematical objects, they are unknowable.

Despite the impact of reasoning like this on a generation of philosophers of mathematics, myself included, I no longer find it compelling. To start, contrary to premiss (1), it is not clear that *any* mathematics is known non-inferentially. Even if some is known directly, I am hard-pressed to think of examples that we would all agree upon. Obviously, those mathematical statements that we know only through proving them do not qualify; so, perhaps, we know the axioms directly. However, many axioms have been proposed, not on the grounds that they can be directly known, but rather because they produce a desired body of previously recognized results. Still others have been developed through a process of trial and error leading from initially inconsistent axiom systems to barely workable repairs, and thence to simple and elegant formulations.[6]

Let us try, then, looking among theorems for directly known truths. Take the commutative law of addition. Since it is so obvious to us now, it looks to be a good candidate for something that *could* be known directly. Yet, can we rule out its being known indirectly? Perhaps it was first inferred from its instances, such as '3 + 5 = 5 + 3'. Or maybe it was somehow inferred from the concept of addition, or someone decided to stipulate that addition be commutative, and found that this worked. What about '3 + 5 = 5 + 3' then? Our mathematical ancestors may have arrived at it by generalizing from their counting experience. And so on.

Perhaps one can put down the doubts we have raised concerning premiss (1). In any case, we can save the argument by dropping premiss (1) and deleting the reference to direct knowledge from premiss

[6] The history of the simple theory of types as it evolved through the work of Frege, Russell, and Ramsey is an example; the course of Quine's New Foundations is another.

(2). So revised (2) amounts to requiring that there be some connection between the knower and the body of facts known, however they are known. (Much of Benacerraf's text supports this interpretation.) Writing in the early 1970s, Benacerraf favoured the then popular causal theory of knowledge as defining the appropriate connection mentioned in premisses (3) and (4). The causal theory of knowledge is no longer in favour, but this need not make Benacerraf's argument obsolete. For it is difficult to see what kind of knowledge- or belief-generating connection, causal or otherwise, there could be between us and mathematical objects.[7]

However, this raises another problem with the argument. For conclusion (7) to follow, one must suppose that the connection between the experiences leading to our knowledge that, say, $3 + 2 = 5$ and $3 + 2$ being 5, which premiss (2) requires, must be established through the numbers themselves. But perhaps the connection is between the facts: we had the experiences we had when we learned that $3 + 2 = 5$ because three plus two is five.[8]

Despite the problems we have found with his argument, the challenge Benacerraf put forth still remains: no matter how strong the prima facie case for mathematical realism, it cannot stand as an ontological doctrine alone. It must be combined with a plausible epistemology. The difficulties we have seen in the argument above simply weaken the case that such an epistemology must be based upon some recognized connection, causal or otherwise, between us and mathematical objects.

But then what constraints should a realist epistemology satisfy? Many epistemologists today hold that a belief cannot count as knowledge unless it has been generated by a process that is reliable. In itself this should be no threat to mathematical realists, since the processes mathematicians use are as reliable as any, given any reasonable construal of reliability. Rather, as Hartry Field has noted,

[7] For a discussion of the causal theory in Benacerraf's argument see Maddy (1990) and Burgess and Rosen (1997).

[8] This objection raises problems of its own. One would want to clarify its use of the word 'because'. Taking it to have the sense of 'if three plus two were not five then we would not have had the relevant experiences' would saddle one with the problem of making sense of conditionals whose antecedents are contrary to mathematics. It might make more sense to maintain that we use 'because' here to indicate that the mathematical truth in question figures in explaining our knowledge of it. For further discussion of the issue of the role of mathematics in explaining our mathematical knowledge see Steiner (1975).

the problem seems to be one of explaining their reliability.[9] For acquiring new beliefs transforms our brains, and whatever processes might lead mathematicians to their beliefs, they cannot physically operate upon or respond to mathematical objects themselves.

Take, for example, the process of multiplying two numbers. We usually operate upon written numerals, but never on the numbers themselves. Why, then, does this produce reliable information about the numbers? Of course, the realist has no explanation to provide if it has to be one that gives mathematical objects some causal or information-transmitting role. But if one doesn't insist on this, there is another perfectly straightforward answer to our question, namely that we multiply by following a sound multiplication algorithm. Assuming that we implement the algorithm correctly, we will not derive, say, '7(43) = 301' unless seven times forty-three is three hundred and one. Now this seems as though I have just pushed the question back a step. The multiplication algorithm is a set of rules for manipulating marks on paper, but the algorithm we prove sound concerns abstract numerals. So now we have two questions: how can manipulating marks on paper, that is, symbol tokens, yield information about abstract numerals, that is, symbol types? And how do facts about abstract numerals reflect facts about numbers? It turns out that we can answer the second question by going over the *mathematical* connections between our system of numerals and the numbers, and the multiplication algorithm and multiplication. This is the same sort of reasoning we use to prove that a system of deduction yields only valid conclusions.

We are left with the relationship between expressions *qua* abstract types and their written tokens. Some find this less problematic than the relationship between numbers and written symbols.[10] But it is not clear to me that it is. After all, we cannot see abstract symbol types any more than we can see numbers. Types are shapes rather than things that have shapes. So why should we expect that looking at shaped marks should tell us about them? Only, I would think, because we suppose that the features of shaped objects reflect or represent the shapes themselves.

Given such a supposition, we can explain why making marks on

[9] Field (1989), 25–7.

[10] Charles Parsons holds such a view, but even he thinks that only a limited amount of mathematical knowledge can be obtained through tokens of numerals. See Parsons (1979–80).

paper can yield reliable information about numbers. In Chapter 9 I will argue that we gain 'access' to mathematical objects by positing them and correlations between some of their features and concrete computations. In Chapter 11 I will develop this idea further by defending the view that the features in question are structural. So in a sense I do hold that we can know types through their tokens. Still, I reject the idea that we can test correlations by directly comparing types and tokens. However, this does not undermine the explanation of why our computations reliably indicate facts about numbers. Even the explanation of why barometers are reliable indicators of weather changes must take some correlations for granted. Barometers indicate weather changes because they register atmospheric changes which are correlated with changes in the weather. We can explain these correlations in turn if we like. But when it comes to explaining the reliability of barometers we must ultimately appeal to a hypothesis that certain features of some device (perhaps a part of the barometer) vary with atmospheric pressure or with something we take to measure it. Of course, we can test our theory of the barometer through using auxiliary hypotheses to deduce observational consequences from it—just as we test other scientific theories. We may be able to test an explanation of our reliability as computers in a similar way, that is, by deducing observational consequences confirming the correlations it uses.[11]

2. HOW CAN WE REFER TO MATHEMATICAL OBJECTS?

Anti-realists have also argued that we cannot refer to mathematical objects because of our lack of causal contact with them. Now to someone unfamiliar with the discussion this must seem a bit puzzling. After all, we can refer to fictions and futures despite our lack of causal contact with either. Why bring causes into a theory of reference?[12]

[11] For further useful discussion of both Field's demand for an explanation of the reliability of our mathematical beliefs and the causal theory of knowledge see Burgess and Rosen (1997).

[12] I direct to Chapter 6 those who are inclined to object that we don't refer to these in the relevant sense, since fictions don't exist and futures don't exist yet. There I give examples of presently existing physical entities with which we can have no causal contact.

Of course, not everybody thinks we should, but one reason people offer for talking about causes in this connection is that we can use them to explain how people who hold very different theories can be speaking about the same subjects. This concern seems relevant to the philosophy of mathematics. We find it plausible that the Greeks knew quite a bit about numbers, and yet some of their ideas about numbers were so different from ours that focusing on our differing ideas might lead us to conclude that their numbers cannot be our numbers. Similarly, we want to say that Newton and Einstein both studied gravitation, despite the radical differences in their gravitation theories.

According to some philosophers of science, we can solve our problem by holding that a given term refers to an object just in case an 'appropriate' causal chain connects the term's users to the object in question. This version of the causal theory of reference permits us to conclude that contemporary biologists speak of the same plants and animals as their predecessors, despite their radically divergent theories, because they all stand in the 'appropriate' causal relations to those plants and animals. Now we need not worry about what 'appropriate' means to see that this theory will not work for causally inert mathematical objects. So the anti-realist can add another strike against mathematical realism.[13]

But must mathematical realists take seriously the problem of divergent views of apparently identical mathematical subjects? Anti-realists need not. If there are no mathematical objects, the only links between our theories and those of other mathematicians can only be logical or linguistic ties between the theories or historical and cultural links connecting the mathematicians themselves. This route is open to mathematical realists too, for as I have characterized realism, mathematical realists are committed only to the truth of *contemporary* mathematics, and not to the truth (or approximate truth) of quite different theories. Thus they can simply say that although there are clear historical links between contemporary and ancient mathematics, there is no more need to decide whether we and our

[13] The causal theory of reference is not retricted to the philosophy of science, but for an influential use of it in this area see Putnam (1973). I should emphasize that the version of the theory I present in this paragraph is an emendation of Putnam's theory that has been popular with mathematical anti-realists. His theory, in which causal chains between *users* of a term determine sameness of reference, is compatible with mathematical realism.

predecessors dealt with the same mathematical realm than there is to decide whether ancient and modern atomists were concerned with the same underlying physical and chemical structures. Furthermore, they could draw support for this from Thomas Kuhn's views in the philosophy of science and Quine's view in the philosophy of language. Both hold that there may be no fact of the matter as to whether people holding sufficiently different theories speak of the same subject-matter.[14]

This hard line fits poorly with some of our pre-systematic realist opinions, however. For it seems to have the consequence of failing to credit our forerunners with much that we think they knew—that Newton knew about gravitation or Euclid of the infinitely many prime numbers. On the other hand, the causal theory of reference is a transcendent theory, and it not only fits poorly with mathematical realism but also with an immanent approach to truth. Despite this, we can find a place for some of the intuitions driving the causal theory by using an immanent, disquotational approach to reference, based on the disquotational biconditionals for names and predicates (Chapter 2, Section 2.3). We must, however, apply the idea to the human polyglot. Then whether we and Euclid are referring to the same thing reduces to the question of whether numbers are *arithmoi*.

This approach does not make the question any easier to answer, but at least it does not rule it out of court at the outset, as the version of a causal theory we have considered seems to do. Of course, in the end, we may conclude that the question has no answer, that there is no fact of the matter in this case. Given that our pre-systematic intuitions often seem to vary with our philosophy, this is not an especially adverse outcome.

3. THE INCOMPLETENESS OF MATHEMATICAL OBJECTS

Puzzles such as those concerning the identity of the Greek numbers, or Newtonian and Einsteinian gravitation, arise because grammar and common sense prompt us to formulate questions even when there may be no conceivable evidence supporting one answer rather

[14] See Kuhn (1977), Quine (1969a).

than another. This situation occurs even more frequently, more clearly, and more urgently in mathematics than it does in the rest of science. It is due to what, following Charles Parsons,[15] I will call the incompleteness of mathematical objects.

Mathematical objects are incomplete in the sense that we have no answers within or without mathematics to questions of whether the objects one mathematical theory discusses are identical to those another treats; whether, for example, geometrical points are real numbers. This springs from the way mathematics defines its terms. When defining bottoms out, that is, when characterizing its primitive concepts, mathematics is content with axiomatic characterizations.[16] This method for dealing with primitives first arose in algebra, where, for example, a group is defined as any set and operation satisfying the so-called axioms of group theory, and it spread through the work of Hilbert and Dedekind to the foundations of geometry and number theory, where Euclidean spaces are often characterized as anything that satisfies the axioms for Euclidean geometry and the natural numbers as anything that satisfies the Peano axioms. As a consequence today mathematics seeks to characterize its objects only 'up to isomorphism', and its most fine-grained axiomatic characterizations—those provided by categorical axiom systems—have infinitely many distinct but structurally identical models. Thus, as far as mathematics goes, the natural number sequence might be one of infinitely many progressions of equally spaced points on a half-line, or progressions of sets, or even progressions of Roman numerals. Neither the axioms nor additional constraints arising from counting exclude any of these alternatives. On the other hand, the natural number sequence cannot be each of these progressions, since some are provably distinct from each other. Yet whether it is one of them or none, mathematics does not say.

Mathematics also reflects the incompleteness of its objects in defining other objects in terms of those taken as primitive. For example, when taking sets as primitive, mathematics authorizes many alternative definitions for its other fundamental objects: thus we have real numbers defined as Dedekind cuts, or alternatively as infinite sequences of rational numbers; natural numbers defined as

[15] Parsons (1979–80).

[16] These are often called *implicit definitions*. Following Frege, I think this term is misleading. For a fuller discussion of Frege's position, see Resnik (1980).

different sets by Frege, von Neumann, and Zermelo; and functions identified alternatively with many–one and one–many relations. Nor need we start out with sets, for an entire mathematical ontology can also be defined in terms of functions, and in terms of ordinal numbers if we limit sets to Gödel's 'constructible' ones.

Now this is no defect in mathematics; leaving these identities unsettled does not hamper its practice or progress. But it does generate a philosophical problem, made famous by a paper of Paul Benacerraf's.[17] Consider this. Zermelo defined the number 2 as the double unit set of the null set, whereas von Neumann defined it as the pair set of the null set and its unit set. The von Neumann and Zermelo two are distinct sets; so the number 2 cannot be identical with both of them. So which one is it? Should there not be a fact of the matter? Should it not be the case that either 2 equals the Zermelo two or it does not? But no mathematical facts decide these questions; identifying 2 with either set, or neither, contravenes no mathematical dictum. Should we not conclude that since the natural numbers cannot be both Zermelo and von Neumann sets, they are not sets after all? By the same token, should we not conclude that the real numbers are not really sequences, infinite sums, or sets of rationals, and that functions are not sets of ordered pairs, and so on? But then by what right would we draw such conclusions? And by what right does mathematics employ the definitions of Cantor, Dedekind, Frege, Zermelo, von Neumann, and others?[18]

These are not puzzles we should take lightly. And one might be tempted to conclude that they constitute an objection to realism, on the grounds that realists should be committed to there being a fact of the matter as to whether, say, numbers are sets, and if so, which sets correspond to which numbers. One might argue that realists take mathematical objects to be things that exist along with ordinary beings, and given any two things, either they are identical or distinct. I grant that this argument does show that realists must deal with the puzzles. Even if they did not appear to pose a special problem for realists, they are important to address. I will also grant, as Hartry

[17] See Benacerraf (1965). This paper also discusses Takeuti's definition of sets in terms of ordinal numbers.

[18] Although Benacerraf's classic statement of the problem of fixing the identity of mathematical objects was restricted to the natural numbers, it should be emphasized that the same problem affects all mathematical objects. See also Kitcher (1978).

Field has pointed out, that it looks as though those who deny the existence of mathematical objects will have an easier time dealing with the puzzles.[19] However, I would add that mathematical realists are not committed to claims about mathematical objects beyond those they hold by virtue of endorsing the claims of mathematics. Since mathematics recognizes no facts of the matter in the puzzling cases, mathematical realists are free to develop solutions that do not recognize them either.

Let us also note that if mathematical realism is in trouble because of the incompleteness of mathematical objects, then other forms of realism are in trouble because of the analogous incompleteness of the objects they recognize. Suppose, to adapt an example of Quine's, that one is a realist about both tables and molecules.[20] According to the reasoning used to generate the puzzles concerning mathematical objects, the table should be identical to some swarm of molecules, but no evidence, physical or otherwise, determines which of the many swarms it is. Or take some person, say Frege, and his body. Some time during its duration this body was identical to Frege himself. But what evidence can determine when Frege began and when his body ceased to be him? Turning to more theoretical entities, we can ask whether ordinary bodies are identical to the space-time regions they occupy or to the mereological sums of their undetached parts, and find no evidence to decide among the alternatives. Thus the incompleteness problem threatens to arise whenever we combine two previously separate universes of discourse into one joint universe, for then we can frame identities that previously did not makes sense.[21]

4. SOME MORALS FOR REALISTS

What morals can we draw from this discussion of the difficulties with realism? First, that we realists must provide an epistemology for mathematics. It need not be based upon some causal connection between us and the mathematical realm, but its account of the generation of mathematical knowledge should recognize only natural

[19] Field (1989), 22–3.

[20] Quine (1981b).

[21] The incompleteness of (mathematical) objects is related to what Quine has called ontological relativity, but it would take us too far afield to sort out exactly how they are related. For further discussion see Resnik (1997).

processes. I don't want to mire in questions about the meaning of 'natural' or the scope of science, so let me simply say that *ideally* the processes in question should be countenanced by established branches of science. However, the processes producing our beliefs need not constitute their justification. This might come later. The view I shall develop in Part Two is that we arrive at our mathematical beliefs by making up mathematical theories. Consequently, it might happen that we do not even believe, much less know, some of the mathematical principles we introduce. Yet this does not preclude their evolving into a body of knowledge once they acquire sufficient supporting evidence.

Second, philosophers of mathematics of every stripe should address the problem of the incompleteness of mathematical and other objects. Solving it for the mathematical case may be easy for those who hold that mathematics is just a system of games or a useful fiction. But it is preferable to seek a more general solution, and here it is not evident that mathematical realists will be at a disadvantage.

Third, realism about mathematical objects does not commit one to transcendent theories of reference, or to the causal theory of reference in particular. Thus the immanent approach I favour still remains an option.

5. AN ASIDE: PENELOPE MADDY'S PERCEIVABLE SETS

Penelope Maddy meets the realist's epistemological challenges by introducing the kind of assumptions I would rather avoid. She attributes physical properties to certain mathematical objects. To begin, she believes that the epistemology of mathematics parallels that of the empirical sciences in that both mathematics and the other sciences support theoretical hypotheses with evidence drawn from a more certain and more elementary database. Her writings have focused on the case of set theory, whose higher axioms, such as the infinity and the power set axioms, she argues, are supported by elementary *observational* knowledge about sets.

Yes, observational knowledge about sets. Maddy breaks with traditional realists by denying that all sets are causally inert and outside of space-time. On her view, sets of located objects are located when and where their members are. Thus with the books on my desk there are also the set of each, the set of any two, the set of any three . . ., the set of them all, the set of any one of these sets, and so on. Indeed, an entire model of Zermelo–Fraenkel set theory exists right where my books are.

Since there are sets in front of us, we might be able to observe them. And Maddy thinks that we can—at least we can see some of the less complicated sets, the set of my books, for example, just as we can see the books themselves. That is why is possible for us to gain observational knowledge about sets. To buttress this she explains how an account of our perception of these sets can be integrated into contemporary perceptual psychology. Now my criticism of Maddy's view will not turn upon the details of the psychological theories to which she appeals, so we need not concern ourselves with them or with their scientific merit. I will grant that if sets are medium-sized physical objects on a par with rows of books, then it would be plausible to regard them as observable. Perhaps the fact that so many people are unaware of the sets right in front of them can be attributed to their ignorance.

Maddy's theory stands and falls with her assumption that certain sets are located along with their members and are no less observable than those members. Now why should we accept this heterodox assumption? Maddy's answer: if one already recognizes the existence of sets, there is no 'real obstacle' to holding that sets formed from spatio-temporal objects are located along with those objects, and that we acquire some beliefs about sets directly through perception.[22]

Now while it may be consistent to hold that sets of physical objects are located in space-time, it is not clear that this makes them into genuine physical objects. For if these sets are physical objects, they should be distinguishable via their physical properties. But the physical sets occupy exactly the same spatio-temporal places as their members and they participate in exactly the same events, so our usual means for distinguishing physical objects do not apply to them. But then what physical properties distinguish a set of two books from the set of this set or the set of non-empty subsets of this set? Maddy would respond that they differ from each other through having different members. But *having different members* is a property

[22] Maddy (1990), 58–63.

of sets, and it is a physical property only by virtue of Maddy's stipulation that these sets are physical objects. Thus appealing to this property begs the question of whether sets are physical.

What about Maddy's claim that we can see certain sets? On her view, when we look into a carton of eggs and observe that there are three eggs in it, we have acquired a perceptual belief about the set of eggs in the carton, namely that it has three members; and, hence, we can learn about sets through direct perception. Now by looking at a computer print-out, trained observers learn things about electrons, genes, or the interiors of stars, none of which can be directly perceived. Thus, contrary to Maddy's inference, acquiring beliefs about sets via perception need not entail perceiving the sets themselves.

Maddy would respond that scientists *infer* their beliefs about electrons and from observation reports and their theories, and that empirical evidence, based upon reaction time studies, shows that we form beliefs, like the belief that the carton contains three eggs, too rapidly to infer them. But this shows little. For it is not at all unusual for a novice to take some time to conclude what an expert can see instantly and automatically. Mechanics and physicians can immediately 'smell' a specific malady in an engine or person. And it is likely that trained scientists can immediately see that some subatomic phenomenon has occurred in a cloud chamber.

Maddy's epistemology does not appeal to the ideal of invoking only processes countenanced by established branches of science. Perceptual psychology does not speak of perception of sets, at least not yet. At best she has offered us a consistent picture of how this branch of science might come to recognize the perception of sets. But it is a picture based upon assumptions and inferences that go against the grain. To review them, first, the only properties that distinguish sets of physical objects from aggregates of their members are properties of sets. These are dubious candidates for physical properties. Secondly, coming to know something about a thing by means of perception does not entail that we can perceive that thing.[23]

[23] For a fuller critical discussion of Maddy's views, see Chihara (1990), Balaguer (1994), Lavine (1992), and Tieszen (1994). I am indebted in part to their criticisms.

PART TWO

Neutral Epistemology

INTRODUCTION TO PART TWO

In the next chapters I begin to develop an epistemology for mathematics, which I will complete in Part Three where I expound my structuralism. The view I present in this part is 'neutral' only in the sense of being largely independent of structuralism. It derives from Quine's epistemology of science, and so in its basic thrust it is postulational and holistic. On this view, we postulate mathematical objects for the purpose of providing science with richer inferential and descriptive methods. To postulate something is simply to make up a theory that asserts that it exists. Doing this is no more occult than writing fiction—a process with which mathematical anti-realists have no quarrel.

Taking mathematical objects as posits removes the mystery of how we came to form mathematical beliefs, but it raises other questions that I must address before my epistemology can be complete. These include the following: What distinguishes mathematics from fiction? Is a postulational epistemology compatible with my realism? What makes our mathematical theories or beliefs *about* mathematical objects?

These questions primarily concern the genesis of our mathematical knowledge. Others concern its justification: what kind of evidence do we have for our mathematical beliefs? What roles do logical deduction, proof, and computation play in justifying these beliefs? How do we justify introducing new mathematical axioms or new kinds of mathematical objects? In answering these questions I will be presupposing a holistic approach to evidence: no claim of theoretical science, including those of mathematics, can be confirmed or refuted in isolation but only as part of a system of hypotheses. Holism fits nicely with a postulational account of the genesis of mathematical knowledge, for it explains why we should not expect to find pre-theoretic evidence for mathematical objects. On my account, *ultimately* our evidence for mathematics and mathematical objects is their usefulness in science and practical life. Thus an indispensability argument is crucial to my epistemology.

A major objection to holism is that the notion of evidence used in science differs markedly from that the holist account of it and from the notion of evidence used in mathematics. Scientists do take experimental results as bearing upon specific hypotheses instead of the entire system of science, and mathematicians virtually never use anything resembling an experiment to test their theories. To answer this objection I will argue that not only is global holism compatible with *overridable* local conceptions of evidence, but there are also pragmatic grounds, based upon the good for science as a whole, for promoting different local conceptions of evidence—especially the local conception of mathematical evidence that differs as it does from that found in the rest of science.

I will also need to address some delicate issues concerning logic, since logical deduction plays a more central role in mathematics than it does in any of the natural sciences. To many, logical deduction remains a solid source of a priori knowledge and necessary truth.

If logical truths are necessary, then they must have some property, besides simply being true, that makes them necessary. And, if we are to be realists, its obtaining must be independent of our conventions, our choice of logical rules, and the like. I shall argue that we have no reason for believing that anything distinguishes so-called logical truths from other truths in this way. In my view, judgements of logicality are normative rather than descriptive.

This is not to say that the statements we call logically true are not true or that we don't know them. (I am not denying the (plain) truth of 'if $0 < 1$, then $0 < 1$ or $0 = 1$', for instance.) So this leaves the possibility of their being known a priori—and they are in a sense. Logic plays such a central role in connecting our theories with experimental evidence that it would be 'illogical' to take some experiment as refuting it. Thus we place our logical truths outside the circuit by which we test our beliefs, thereby allowing no experience to confirm or refute them. Yet if, by our lights, our system of beliefs fails to fit experience, we have the option of revising our notion of fitting experience—even to the extent of revising the limits of logic, altering our inferential norms, and in so doing changing what we count as logically true. Thus even logic is not a priori in the sense of being immune to revision in the light of experience.

6

The Elusive Distinction between Mathematics and Natural Science

I will mark no sharp epistemic distinction between mathematics and the rest of science. Part of my reason for taking this course is that I know of no viable way of drawing such a distinction that is also open to realists. Truth manufactured by convention is no option for them. Frege's programme for reducing mathematics to logic is not technically convincing, and, in any case, it depends upon realism about logical truth, which I reject. Gödelean intuition is too mysterious, and so on. I need not belabour these considerations; they are familiar points in the philosophy of mathematics.

Still it is surprising how the view persists that the epistemology of mathematics must be different in kind from the epistemology of empirical science. One of the major reasons for this, I think, is that most philosophers think that mathematical objects differ metaphysically from the objects the other sciences study. They have no location in space-time; they are causally inert and experimentally undetectable. I agree that mathematical objects have these features, and I also believe that most of the objects the other sciences study do not share them. But enough physical objects do share them to break down the epistemic and ontic barriers between mathematics and the rest of science. I shall argue this by examining the ontology of theoretical physics.[1]

[1] Few have questioned the ontic distinction between mathematics and science that so much philosophy of mathematics presupposes. LaVerne Shelton challenged the abstract–concrete distinction in an unpublished address to the American Philosophical Association (Shelton (1980)). Susan Hale continued this line in her dissertation and related articles (Hale (1988a, 1988b)). Hartry Field's taking space-time points and regions as a concrete, ontological foundation for his nominalization of physics prompted me to criticize his presupposition: Resnik (1985a).

1. HOW PHYSICS BLURS THE MATHEMATICAL/PHYSICAL DISTINCTION

Most anti-realists in recent philosophy of mathematics have been realists about theoretical physics. Typically, they argue their anti-realism by emphasizing that mathematical objects (if there are any) present different and more intractable epistemological problems than those introduced in physics, since the latter are in principle detectable due to their being located in space-time and being causally active. In the next sections I will employ examples from contemporary physics and cosmology to undercut the supposed sharp distinction between mathematical and physical objects that underlies contemporary thinking about the epistemology of mathematics. Of course, anti-realists in the philosophy of science may well object to these examples as fictitious, but, this is not an option for most anti-realists in the philosophy of mathematics.

1.1. *Quantum Particles*

Let me start by considering some physical objects that appear to be as much mathematical as physical. I have in mind quantum particles. The term 'particle' brings to mind the image of a tiny object located in space-time. But, on what seems to be the consensus view of the puzzling entities of quantum physics, this image will not do. Most quantum particles do not have definite locations, masses, velocities, spin, or other physical properties most of the time. Quantum mechanics allows us to calculate the probability that a particle of a given type has a given 'observable' property, such as having a position, momentum, or spin in a given direction. (Although texts use the term 'observable', one needs instruments to determine the presence of these properties.) But this does not imply even that if we, say, detect a photon in a given region of space-time, then the photon occupied that position prior to our attempts to detect it or that the photon would have been in that region even if we had not attempted to detect it. Prior to its detection a photon is typically in a state that is technically called a superposition of definite (or pure) states, and quantum theory contains no explanation of how a photon or any other quantum system goes from a superposition into a definite state. Recent mathematical critiques of hidden variable theories indicate that this mysterious feature of quantum mechanics is virtually unavoidable.[2] In spite of this, one might still say that quantum particles are tiny bits of matter with very weird properties—ones that are only partially analogous to classical physical properties. But there is a further problem.

When we have a system of several particles of the same kind—two photons, for example—there are no quantum-mechanical means for tagging them at the beginning of an interaction and re-identifying them at the end of it. Suppose, for example, we run an experiment in which we detect two photons one above the other at one location and later detect two photons, again one above the other, at a location a bit to the right of the first. Intuitively, they are the same two photons, and one would expect that there is a fact as to whether the photon on top at the beginning is also on top at the end of the experiment. But quantum theory recognizes no such fact; for such a fact must be described by reference to the trajectories the particles take in travelling between the two locations. Since their initial and final positions in space-time are fixed, they have definite trajectories only if they have definite velocities, violating the famous uncertainty principle of quantum mechanics. The situation is even worse when we combine quantum theory with special relativity, for then there is no fact as to whether we even have the same two photons. It might be, for example, that one photon splits into an electron–positron pair, whose members in turn annihilate one another and produce the photon we detect. Moreover, between our two detections such processes might occur indefinitely many times. Actually, the same is true of experiments 'tracking' a single photon. The one with which we start might be destroyed before we see the one with which we end.

This means that quantum particles are incomplete in an even more radical sense than mathematical objects. For in mathematics there may be no fact of the matter as to whether mathematical objects treated by different branches of mathematics are the same or different—whether, for example, a given real number is a given set—but in quantum physics there may be gaps concerning the identities of objects within the same universe of discourse.

One way to avoid this incompleteness is to think of particles as features of space-time—more like fields—rather than as bodies

[2] See Shimony (1989).

travelling through space-time. This view is seconded in the following passage from a recent account of particle physics directed at other scientists:

> In the most sophisticated form of quantum theory, all entities are described by fields. Just as the photon is most obviously a manifestation of the electromagnetic field, so too is an electron taken to be a manifestation of an electron field . . . Any one individual electron wavefront may be thought of as a particular frequency excitation of the field and may be localized to a greater or lesser extent dependent upon its interaction.[3]

Now by thinking of the behaviour of iron filings around a magnet we can get an intuitive grasp of what a magnetic field is, how it has a source, and an intensity or direction at a given point. However, quantum fields are not distributions of physical forces; rather they are, roughly, distributions of probabilities. As the electron field varies its intensity over space-time so does the probability of an interaction involving electrons.[4] And remember, we cannot think of the electrons as definitely present prior to any electron interaction we observe. Nor are these probabilities merely statistical. (When we measure the strength of an ordinary magnetic field, for example, there is some probability of failing to get a theoretically predicted true value. But this probability can be construed statistically.) In the quantum fields the probabilities must be regarded as irreducible features of the fields themselves (or of space-time).

This suggests to me that quantum fields straddle the border between mathematics and physics. Under certain conditions they have 'observable' physical properties, under others they are little different from functions from space-time to probabilities. 'But,' one will object, 'surely, fields are not the functions themselves; the functions simply represent their behaviour.' Now this ploy might be open to those taking an anti-realist stance towards quantum fields, but my concern is with realists. On their view, a field is real even when in a superposition of observable states. The realist objectors are committed to the superposition's being a physical substance. Let them use the term 'physical' here. I am not claiming that these entities are definitely mathematical. But let us not let the term 'physical' obscure how unlike ordinary physical objects (or ordinary fields) quantum

[3] See Dood (1984), also Teller (1990).
[4] Cf. Teller (1990), 612–13.

fields in superpositions are. Remember, they have manifestations, but they don't cause them. And some, such as superpositions of space-time position, may become localized or take on a definite orientation, but prior to that they 'occupy' all of space-time or have every orientation.

David Bohm's interpretation of quantum physics lends credibility to viewing quantum objects as quasi-mathematical.[5] To try to make sense of the paradoxical features of quantum particles, Bohm proposed taking a particle's wave function as physically significant instead of a mere mathematical means for representing the particle's behaviour. It is to be a force field of sorts that guides the particle from one place to another. But where the usual theory treats the particle as in a superposition, Bohm's theory interprets the wave function (now viewed as a force field) as splitting into parts with only one part accompanying the particle. *The remaining parts of the wave function/force field are completely undetectable, are causally inert, and have no effects on other particles.*[6] Furthermore, although there is a fact of the matter as to which part of the wave function accompanies the particle, and hence as to which trajectory the particle takes, no physical evidence will reveal these facts to observers. Bohm's proposal blurs the distinction between mathematical and physical objects, because the vacant parts of wave function are undetectable and causally inert. (But presumably they are located in space-time and thus not fully abstract.)

Making sense of the paradoxical features of quantum mechanics has motivated much work in the philosophy of physics, and has driven several prominent physicists and philosophers of science towards anti-realist views of the theory. But our concern is with those who conjoin realism about science with anti-realism about mathematics. Quantum mechanics challenges them to explain quantum particles in uncontroversially physical terms.

How might they respond? One approach would be to identify probabilities with propensities (of quantum systems to yield certain kinds of interactions) and quantum fields with distributions of propensities, and then argue that both the distributions and the propensities themselves are physical entities. But it is unclear to me

[5] I rely here on David Albert's exposition of Bohm's views. See Albert (1994).
[6] Ibid. 66.

how the case that these are physically real could be any more compelling than the case for the physical reality of quantum wave functions or quantum fields. Either case involves sacrificing the kinds of properties and causal powers we usually think are essential to physical objects.[7]

1.2. *Undetectable Physical Objects*

According to most philosophers of mathematics, the major epistemic difference between physical and mathematical objects is that the latter cannot participate in causal processes that would permit us to detect them. It is pertinent, then, that physicists now recognize physical entities suffering from the very epistemic disability ascribed to mathematical objects. Take, for example, the photon–electron–positron–photon transformations we discussed earlier. Quantum field theory posits processes of this type that happen so fast that they are in principle undetectable. Virtual processes, as they are called, need not even conserve energy or momentum so long as the total processes of which they are components do. If physicists eventually adopt Bohm's interpretation of quantum mechanics, then most physical objects will be empty, physically undetectable wave fronts! For a final example, let us turn to cosmology and the interiors of black holes in space-time. According to physicist Clifford Will, 'there is no way for any external observer to determine, for example, the total number of baryons [inside a black hole]' and 'there must exist [such] a singularity of space-time at which the path or world line of an observer who hits it must terminate, and physics as we know it must break down'.[8]

I would expect someone to reply at this point that, unlike mathe-

[7] Compare this with R. I. G. Hughes's discussion of his interpretation of quantum mechanics via 'latencies':

> The properties of classical systems were summarized by its state . . . The latencies of quantum physics are also represented by the state—here the state vector. The latencies assign probabilities to measurement outcomes . . . A quantum measurement should be regarded neither as revealing a property of the system nor as creating that property, for the simple reason that quantum systems do not have properties . . . For, as Born pointed out, the 'waves' . . . are 'probability waves' . . . Similarly, to ascribe a latency to a system with respect to its position is just to say that there is an extended region of space within which there is a nonzero probability of finding it. (Hughes (1989), 302–3.)

[8] Will (1989).

matical objects, the interiors of black holes, virtual processes, Bohm's wave fronts, and other examples that one might cite are supposed to be part of the spatio-temporal causal network. Yes, virtual processes are supposed to be physically real processes that happen too rapidly to detect, but it is not clear that the interiors of black holes or the vacant parts of Bohm's wave fronts are supposed to be physical in any ordinary sense.[9]

'Well, aren't empirical theories committed to them?', the objector might continue. Yes, but the same theories are also committed to mathematical objects. More to the point, these examples show that combining mathematical principles with empirical hypotheses can commit one to objects whose status is neither clearly mathematical nor clearly physical.

2. SOME OTHER ATTEMPTS TO DISTINGUISH MATHEMATICAL FROM PHYSICAL OBJECTS

We began by observing that it is usual for philosophers of mathematics to distinguish between supposedly abstract, mathematical objects and supposedly concrete, physical ones by appealing to the spatio-temporal locatability, causal powers, or detectability of the latter. We now see that distinguishing between abstract and concrete objects in this way is obsolete. Contemporary physics is willing to entertain theories positing entities that might be undetectable or not in space-time (in any ordinary sense) or causally inert. Still, I can think of other ways in which one might distinguish the mathematical from the physical, and I would like to make some brief points against them.

One of these ways is to discriminate between the physical and the mathematical by claiming that mathematical objects cannot change their properties. But this will not stand up to a first set of objections

[9] These objects are not what Jody Azzouni calls *thick* posits, that is, we don't need 'to interact causally with them in some way to guarantee that they are empirically real'. Rather, on Azzouni's view they would be *thin* posits, whose kind includes certain abstract entities. See Azzouni (1994), 65–6, 74. He and I differ concerning mathematical objects, however, all of which he calls *ultrathin* posits. To continue to put the matter in his terms, I will maintain in Chapter 8 that initially our epistemic practices treated mathematical objects as *thin*—part of their justification was their success in applications—but our practices have developed to the point where we are willing to entertain mathematical posits which we treat as *ultrathin*.

either. Numbers do change some of their properties. Just as Smith may be thin as a child, and not as an adult, the number 60 may register Smith's height in inches at age 12 and not at a later age. Now one might object that if we take the property in question to be *registering Smith's height at age 12* then it is an eternal property of 60 instead of a temporary one. But if we must add a time specification here, then we should also do so for other properties, such as the property of being *thin*. Yet if we replace *thin* with *thin at t* for an appropriate time *t*, then Smith will not change properties either. Perhaps one will object that *registering Smith's height at age 12* should not count, because it is just an accidental property of a number. But physical objects change only their accidental properties too. Perhaps the property of *registering Smith's height in inches* should not count because it is not a real property. But what makes it unreal? Surely not simply because it is a relational property, since electrons, say, have properties only by virtue of their relations to other particles.

Similar difficulties surround the proposal that no mathematical object can participate in an event while every physical object can. If to participate in an event is to be involved in a causal interaction constituting all or part of the event, then we have already seen reasons for doubting that certain supposedly physical objects can participate in events. On the other hand, if to participate in an event is to change one's properties during the course of the event, then the numbers and functions used to characterize physical events do change their properties during such events. Now the most persuasive way around this point would be to show that every event can be fully and precisely described without referring to mathematical objects. We know that we can do this in some elementary cases, but we have little reason to think that we can do this with events involving subatomic particles, whose basic features do not correspond to anything uncontroversially physical.

3. OUR EPISTEMIC ACCESS TO SPACE-TIME POINTS

The acceptance of Einstein's general theory of relativity has led many philosophers to regard the geometry of space-time as an empirical theory and space-time itself as a physical entity. Capitalizing on this, a number of philosophers of mathematics, in particular both Hartry Field and Geoffrey Hellman, have declared it acceptable for mathematical anti-realists to employ space-time points in their constructions.[10] Now whether space-time points are mathematical or physical, abstract or concrete, there will be no real gain in using them to dispense with (other) mathematical objects unless they are more epistemically accessible than the objects they replace.

Hartry Field gives us reasons for thinking that they are. First, we can observationally test theories about the structure of space-time, as we have tested the general theory of relativity. Second, space-time points and regions are part of space-time, and thus part of our physical world. Some, he says, even fall within our field of vision. Finally, if we take the eminently plausible step of identifying fields with properties of space-time, then space-time points and regions become causal agents.[11]

Let us consider Field's reasons one at a time. I grant that we can observationally test theories about the structure of space-time. But we do so by virtue of testing larger bodies of theory that contain a space-time theory, and by holding various auxiliary hypotheses fixed. Furthermore, we can apply such indirect tests only to some geometric hypothesis, such as those concerning the metric of space-time. In particular, I know of no way of testing the claim that space-time points exist. What is more, if these indirect tests are taken to show that space-time theory is empirically testable, then the same can be said for certain mathematical theories. For example, with the appropriate set of auxiliary hypotheses we can use computer runs to test certain claims about Turing machines. (I will say more about these issues in the next two chapters.) So far, then, there seems to be no epistemically significant difference between points and numbers.

But, according to Field, points and regions fall within our field of vision. I am not sure what he means when he says this. He may just mean that they are located in front of our eyes. Of course, that does not imply that we can see them, any more than it does in the case of microscopic bits of matter, virtual processes, or Bohm's wave fronts. On the other hand, he may mean that points and regions are among the things we could see if we paid enough attention, knew what to look for, and so on. I doubt he meant to include points or tiny regions. So his thought must have been that we can see medium-sized regions and thereby ground a theory which posits smaller ones

[10] Field (1980), Hellman (1989).

[11] Field (1989), pp. 68–70.

and points. If we see regions, we do so because the matter in them reflects light. But if instead of attributing this ability to matter, we attribute it directly to portions of space-time, then we see regions because of their causal powers. So the cogency of Field's second reason stands and falls with his third, his attribution of causal powers to points and regions.

Suppose, then, that we attribute the causal powers of bits of matter and physical fields to the space-time regions they occupy. Then our epistemic access to certain medium-sized regions of space-time is unproblematic. But that still leaves points and smaller regions. We should not conclude that the accessibility of the larger transfers to the smaller. After all, undetectable virtual processes differ from ordinary quantum processes only in their brevity.

What, then, gives us our epistemic access to points and tiny regions? I think we should distinguish two questions here: (1) How did we come to know about points? (2) Given what we know now, how might we detect them? The answer to (1) is that we inherited a physical framework positing points from our scientific ancestors. They introduced them in the course of describing the continuity of space and time. We have accepted this hypothesis, because it was and remains the simplest way of formulating our physics and accounting for our informal experience of space and time. (Things seem to move continuously through space.) As to (2), given what we know now, I can think of no reason why we should be able to detect points or very small space-time regions (as opposed to larger regions containing them).

4. MORALS FOR THE EPISTEMOLOGY OF MATHEMATICS

I will not go so far as to claim that the previous sections constitute fatal blows to the distinctions between mathematical and physical objects that I have considered, but they do show that some careful work must be done before we can take these distinctions for granted. If this is correct, then much contemporary philosophy of mathematics has been based upon a dubious and poorly defended presupposition. Nominalists and other anti-realists should not dismiss realism on the grounds that the epistemology of objects posited by physics will automatically be less problematic than that for mathematical objects. Nor should we realists assume that our epistemology must be radically different from ordinary scientific epistemology.

The epistemology of mathematics that I shall develop over the next few chapters is quite similar to the epistemology of space-time points and many other physical objects. Our mathematical predecessors in antiquity posited numbers in the course of developing a framework for making sense of their experience with counting, measuring, and record-keeping. And, as with space-time points, we can empirically test certain hypotheses about numbers—modulo an appropriate set of auxiliary hypotheses—but the bulk of our evidence for numbers has been the success of the framework positing them.

7

Holism: Evidence in Science and Mathematics

I want to set aside for now the question of how we generate our mathematical beliefs and theories, and focus on questions concerning their justification. In separating questions of justification from those of genesis I obviously do not plan to argue that simply going through the processes that lead us to mathematical beliefs is already enough to justify them. Since I have already announced that my account of the genesis of our mathematical knowledge will be postulational, it should come as no surprise that I will opt for an indirect theory of justification. It is not inconsistent to hold that we first hypothesize mathematical entities and then seek (and sometimes find) direct evidence for them—say, in the form of mathematical intuition. But I suspect that anyone who believes that we can know mathematical objects directly would find the idea that we first postulate these very objects both unnatural and extraneous. (It would not be unnatural to maintain—as, indeed, Maddy has—that we know some mathematical objects (for example, finite sets) directly and posit others (for example, infinite sets).

Of course, we justify many of our mathematical beliefs by proving them. But plainly this is only part of the story of mathematical justification, since proofs can carry no more justificatory weight than the premisses—the axioms—upon which they rest. Once one makes this observation, it is hard to resist the temptation to narrow the question of justification to that of the axioms and try to develop a foundational account of mathematical knowledge. We should resist this temptation. Historically many, if not most, branches of mathematics have become settled bodies of knowledge prior to their being axiomatized. When mathematicians have axiomatized them they have done so more with an eye to organizing and systematizing them than justifying them. Moreover, in selecting their axioms they have been more concerned with obtaining elegant, minimal, and independent systems than with providing epistemic foundations.

You might find these historical considerations irrelevant, if you thought that you could justify various branches of mathematics by reconstructing their foundations.[1] But I find it hard to have much confidence in foundational approaches to the epistemology of mathematics. Logicism did not work.[2] *Truth* by convention will not work for realists, for whom truth is independent of our theorizing. But what about *justification* by convention or stipulation, which takes the axioms as justified by stipulation? If this simply means that axioms are like hypotheses and revisable if they yield unwanted conclusions, then we do not have foundationalism. If it means that they are like clauses in a definition, defining the mathematical framework to which they belong, then we are still up against the problem of finding evidence that our definitions are not vacuous. This is not something we can make true by definition, even if the only condition we require for being justified in believing a set of axioms *simpliciter* is that we are justified in believing them consistent.[3]

Consistency proofs can help, but ultimately they transfer the demand for evidence to some other system, and we would need to break this cycle. According to a traditional view, we can give a non-mathematical demonstration of the consistency of a system by exhibiting a so-called concrete model for it. But simply pointing to some part of the physical world is not enough; we must characterize it in sentential terms. So, in the end, the method of concrete models amounts to translating mathematical sentences into physical terms. Philosophers of science often illustrate this by articulating a geometry in terms of light rays. The success of the resulting optical theory is supposed to be evidence for the consistency of the geometry. Of course, if this is the way we ultimately establish the consistency of

[1] Cf. Mark Steiner's discussion of the logician midwife in Steiner (1975).

[2] Some philosophers have argued that logicism will work for second-order number theory. But it is debatable whether the axioms they have proposed are logical or analytic, and, in any case, the resulting system still falls short of supplying the mathematical needs of science. See Wright (1983) and Resnik (1984).

[3] At times Hilbert stated such a view, and lately Mark Balaguer has. See Hilbert (1971) and Balaguer (1995).

our mathematical theories, then our evidence is ultimately empirical and the epistemology of mathematics is non-foundational.[4]

We sometimes justify believing in (the truth or consistency of) some mathematical theories by appealing to others. But because I know of no convincing theory of how to bottom this out in direct justifications, my own view is that some mathematics must be justified indirectly in terms of its consequences. This raises the question of what counts as a relevant consequence. One might see mathematics as a totally distinct science that parallels the rest of science, as Penelope Maddy has, or one might see mathematics as part of science. If we take the former course, then the consequences must be mathematical. What, then, shall we count as the evidential database? Elementary truths of geometry, arithmetic, and set theory? Perhaps, but how then are these known? Maddy has argued that we know the elementary truths of set theory by intuition, but as we noted in Chapter 5 the recent literature has been highly critical of her view. It is unlikely that one can give any better account by focusing instead on numbers or geometric figures.

The idea that the data immediately supporting mathematical theories are such elementary truths seems right, but it also seems to me that these truths are supported indirectly themselves via their connection with so-called empirical truths. In this chapter and the next I shall try to make the case for this by defending an epistemological holism that includes both mathematics and the other sciences within its scope.

1. THE INITIAL CASE FOR HOLISM

By holism I shall mean epistemic or confirmational holism, that is, the thesis that no claim of theoretical science can be confirmed or refuted in isolation but only as part of a system of hypotheses. The

[4] The method of concrete models raises a number of complicated questions. Here are two that occurred to me: (1) Must the physical theory be true or need it only be predictively successful? (2) As we learned from the example of general relativity theory, physical theories or methods may already embody the very mathematical assumptions whose consistency may be in question. Does this change the epistemic situation? Given the position I will develop in this chapter and next, I am inclined to say 'No' to (2), and to say 'The theory need only be predictively successful' to (1).

nub of the argument for holism consists of an observation about science and a simple point of logic.

The observation about science is simply that the statements of any branch of theoretical science rarely logically imply observational claims when taken by themselves, but do so only in conjunction with certain other 'auxiliary' hypotheses. Thus, for example, taken in isolation the statement that gasoline and water do not mix does not imply that when I combine water and gasoline in a container I will be able to observe them separate. It only does so under the assumption that the container contains no other chemical that allows them to homogenize, that the container is sufficiently transparent, that my eyes are working, and so on. Hence—and this is the point of logic—if a hypothesis H only implies an observational claim O when conjoined with auxiliary assumptions A, then we cannot *deductively infer* the falsity of H from that of O but only that of the conjunction of H and A, $H \& A$. Furthermore, if we subscribe to a confirmation theory, on which a set of hypotheses is confirmed by its true observational consequences, then the truth of O confirms not H but rather $H \& A$. Strictly speaking, it is systems of hypotheses rather than individual ones to which the usual, *deductively characterized*, notions of empirical content, confirmation, and falsification should be applied.[5]

The previous paragraph has its roots in Pierre Duhem's writings. Duhem also defended the law of inertia and similar physical hypotheses against the charges that they have no empirical content and are unfalsifiable. One way of putting the law of inertia, you will recall, is to say that a body remains at rest unless an external force is imposed upon it. Since we can only determine whether something is

[5] I don't intend to delve into confirmation theory here, but some brief remarks may be in order. The (Duhemian) points in the text apply to confirmation theories which maintain that unless a system of hypotheses S implies an observation statement O, O cannot confirm S (or belong to its empirical content) and the negation of O cannot falsify S. I would think that something like the Duhemian points would also apply to probabilistic approaches to confirmation, because the same considerations that show that auxiliary hypotheses are necessary for forging deductive ties between theory and observation are likely to apply to establishing probablistic relevance. An exception to this is a Bayesian approach allowing agents to start with arbitrary probability functions. Since both Bayesian and other probabilistic approaches take mathematics as part of the background framework—the 'underlying logic'—in which confirmation is defined, Quine's extension of Duhem's reasoning (see below) does not apply to them.

at rest by positing some observable reference system, this law, taken by itself, implies no observational claims. Furthermore, by appropriately changing reference systems we can guarantee that a body moving relative to our present system is at rest relative to the new one, and thereby protect the law against falsifying instances. All this troubled the law's critics, because they believed that as a physical law it should have an empirical content and be falsifiable. Duhem responded to their worry by observing that the law readily produces empirical consequences when conjoined with auxiliary hypotheses fixing an inertial system; and that in needing auxiliaries to produce empirical consequences, it was no different from many other theoretical principles of science, whose empirical content everyone readily acknowledged. Thus the law's critics could not have it both ways: to the extent that their critique challenged the empirical status of the law of inertia it also challenged that of most other theoretical hypotheses.[6]

Using logic to extract observational consequences from the law of inertia also depends upon including mathematical principles among the auxiliary hypotheses. Duhem drew no conclusions from this about mathematics. But Quine subsequently did. Using the very strategy Duhem used in defending the law of inertia, he argued that even mathematical principles, which by most accounts are just as unfalsifiable and devoid of empirical content as the law of inertia, share in the empirical content of systems of hypotheses containing them.[7]

I do not know why Duhem did not come to the same conclusion as Quine did. However, Frege, who anticipated Duhem's remarks on the law of inertia,[8] would have had a reason for not doing so. For him, mathematical principles are logical principles, and as such implied by the conjunction of the law and its physical auxiliaries. Thus, to him, it would be a mistake to hold that deducing observational consequences from the law depends upon including mathematical principles among the auxiliaries, and also a mistake to conclude that failed predictions tell against a system of hypotheses containing mathematical principles.

Quine would not accept this conclusion, because he restricts logical implication to that afforded by first-order logic with identity.

[6] Duhem (1954). [7] Quine (1990), 14–15. [8] Frege (1891).

Still he recognizes at least this much of Frege's point: whatever the limits of logic, systems of logical truths are immune from the direct sort of empirical falsification to which systems of mathematical and scientific hypotheses are subject. This emerges in the following passage from one of his recent books:

> Now some one or more of the sentences in *S* are going to have to be rescinded. We exempt some members of *S* from this threat on determining that the fateful implication still holds without their help. Any purely logical truth is thus exempted, since it adds nothing to what *S* would logically imply anyway.[9]

In earlier writings Quine even suggested that we might revise logic to save a theory in the face of contrary experience.[10] At first sight this seems to conflict with the passage just quoted, but in correspondence Quine explained that when we revise logic to save a hypothesis in the face of conflicting experience we effectively refuse to acknowledge the 'fateful implication' as such.

Now one might wonder how revising logic could even be an option for us. For without logic the experience would be neither connected to the theory nor contrary to it. It is true that without some logical framework, hypothesis testing could not take place, but that does not mean that the framework and the hypotheses tested cannot both be provisional. Obviously, revisions in the framework must come very gradually, since after changing it we will need to check on whether previously tested hypotheses still pass muster. Thus instead of denying all instances of, say, the law for distributing conjunction over alternation, we might reject certain applications of it to quantum phenomena. In this way there would be no danger of lapsing into total incoherence. Nor need we abandon the norms surrounding deduction. While we may change, for example, what counts as an implication or a contrary, we need not abandon norms that commit us to what our theories imply or that prohibit us from simultaneously maintaining two contraries.

Still, even if we allow revisions in logic, it enjoys a special methodological status by virtue of fixing the very framework through which hypothesis testing takes place. We shall see that because of the prominent place of deduction in its methodology, mathematics shares the methodological halo with logic, which

[9] Quine (1990), 14. [10] For example, Quine (1951)

makes it easier to protect from empirical refutation than the rest of science.

2. OBJECTIONS TO HOLISM

The prominent place of deduction in the methodology of mathematics also reinforces a form of objection to Quine's holism that one frequently encounters, namely that it fails to respect certain intuitions about mathematics and logic that seem to be firmly supported by our mathematical and scientific experience. According to these intuitions, mathematics and logic are fixed points in our investigation of the world, determining the limits of what we can entertain as serious possibilities (to borrow a phrase from Isaac Levi).[11] In a different vein, Charles Chihara and Charles Parsons have objected to the concept of mathematical evidence that appears implicit in Quine's views on theory-acceptance confirmation.[12] On the one hand, Chihara observes that in deciding whether to add a new axiom to set theory no set theorist is going to investigate its benefits to the rest of science.[13] On the other hand, Parsons has pointed out that Quine's holistic picture of science seems to belie the intuitively clear separation between mathematics and the rest of science and to provide no place for specific kinds of mathematical evidence that many believe we have.[14]

In addition to this, some philosophers continue to hold, contrary to both Quine and Duhem, that observational evidence can be seen to bear upon specific hypotheses instead of whole systems. Clark Glymour, for example, argues that by taking confirmation as a three-termed relation between the evidence E, a hypothesis H, and a theory T, we can see that for a fixed T, E is relevant to some H in T and not to others.[15] Strictly speaking, Glymour is not denying Duhem's and Quine's point about a two-termed confirmation relation but rather is taking an approach to confirmation that takes account of their ideas. Nor does Glymour's view imply that mathematical principles are never confirmed or confirmable by empirical

[11] Levi (1980). [12] Parsons (1986). [13] Chihara (1990), 15.

[14] Parsons (1986), 380–3. He also presses these points quite forcefully in Parsons (1979–80).

[15] Glymour (1980).

evidence. Indeed, by holding fixed a certain empirical theory of a physical system exhibiting a given mathematical structure, we might experimentally test certain mathematical conjectures concerning the structure.

On the other hand, Elliott Sober explicitly directs his separatist view of confirmation against Quine's philosophy of mathematics (and by implication Duhem's philosophy of science).[16] Scientists design tests, he observes, to decide between competing hypotheses. They intend to put specific hypotheses at risk, and consequently take the data to reflect upon just these hypotheses and not upon the broader systems to which they belong.

Sober also notes that scientific tests never, or hardly ever, put mathematical claims at the risk of being falsified. Because of this, he argues, mathematics cannot share in the confirmation afforded to those hypotheses that do pass such tests. In particular, the mathematical theory of sets, in contrast to, say, the atomic hypothesis, cannot claim empirical support. He concludes that the confirmational indispensability argument based upon holism fails.

Sober's remarks about scientific practice are certainly right. Neither scientists nor mathematicians take scientific experiments as providing evidence for the mathematics used in designing them. Yet, as I will argue below, even within a holist framework one can make sense of the scientific and mathematical practice to which Sober, Chihara, and Parsons have called our attention.

Before I proceed to this and other matters let me comment briefly on Sober's case against confirmational holism. I take it that his point in describing scientific practice is to exhibit rationally defensible principles of methodology and experimental design that are supposed to conflict with holism. I grant the rationality of the practice Sober indicates, but I deny that it refutes holism. Holists may readily admit that it is rational for scientists to fix certain hypotheses (as auxiliaries) while testing others, and thus also rational (in the practical sense) for them to act *as if* the evidence they obtain bears upon the specific hypotheses being tested. Holists simply deny that, independently of holding the 'auxiliaries' fixed, a logical (or a priori) relationship obtains between the hypotheses tested and the evidence. As Duhem put it, 'these reasons of good sense [for favouring certain

[16] Sober (1993).

hypotheses] do not impose themselves with same implacable rigor that the prescriptions of logic do.'[17] I also doubt that Sober's account of confirmation applies to all non-mathematical hypotheses, and for the sorts of reasons that Duhem emphasized. Specifically, I doubt that many scientific framework and conservation principles—such as the continuity of space-time or the conservation of mass-energy—can be put to the specific sorts of tests that Sober has in mind. Yet we do not want to be forced to deny them empirical content or to hold that the general theories containing them have not been tested experimentally.

The Duhemian argument for holism transfers the burden of proof to those who deny holism. I will call them *separatists*. Unless they can refute Duhem's point of logic or his observation concerning theoretical hypotheses, they must show that the relations obtaining between specific statements and sensory experience that holists attribute to 'good sense' are backed by evidential relations holding independently of our judgements of 'good sense'.

Of course, the same goes for Quine's extension of holism to mathematics. At the time Quine proposed this extension, many philosophers believed that the logicists' identification of mathematics with logic or the positivists' conventionalist doctrine of mathematical truth could answer Quine and successfully account for the objective apriority of mathematics. Largely as result of Quine's criticisms of logicism and conventionalism, few philosophers today believe that we now have a successful account of the apriority of mathematics. Despite this, many remain convinced that, due to its distinctive methodology and its special role in science, mathematics must be a priori. I will not review Quine's criticisms here, for they are well known, and rehearsing them will not change contemporary separatist attitudes. Instead I will show that 'good sense', in the form of pragmatic rationality, underwrites the special role mathematics has come to play in science and bids us to treat it *as if* it were known a priori. This will undercut the methodologically grounded arguments in favour of the apriority of mathematics.

[17] Duhem (1954), 217–18. In view of this remark, I would think that Duhem would find Glymour's ideas a congenial way of working out some of the principles good sense endorses.

3. TESTING SCIENTIFIC AND MATHEMATICAL MODELS

Before proceeding any further it will be useful for us to have a more detailed picture of the connection between scientific and mathematical theories and the evidence. The view that I favour is one developed with great precision by Henry Kyburg.[18] On this view, in science, engineering, and practical life we use combinations of mathematical and scientific principles to develop models (mini-theories) that allow us to calculate values or ranges of values of the quantities that interest us. Then we compare the values we have calculated with the data we have obtained independently (or had independently expected to obtain) in order to decide whether the models are good enough for our purposes. (In science, statistical techniques often guide these decisions.)

Rather than use Kyburg's technical approach, I will explain what I have in mind with an example. Suppose that I want to cross Long Island Sound in a small sailing boat equipped with a compass and navigational charts. If I am near the New York City end of the Sound and the day is clear, I will be able to see from one side to the other. In this case I will not worry much about plotting a course, much less the best course I can. But if it is foggy or late in the day or I am at the other, wider end of the Sound, I will not have that luxury. The simplest thing I might do is to derive a compass heading by 'walking' a straight line drawn on my chart over to the chart's compass rose.[19] I will be satisfied with this approach if I am confident that it will bring me close enough to my destination. On the other hand, I may worry about periodic variations in the Earth's magnetic field and try to determine how much to swing my straight line course to compensate for it. Or I might plot a jagged line instead of a straight one in order to compensate for the tides and the winds. Whatever I do, it is likely that I will make certain simplifying assumptions (for example, I will almost certainly use an average wind velocity and probably an average current velocity). Thus the most I can conclude is that, with a certain degree of confidence, the

[18] Kyburg (1990).

[19] At a first pass we might take my 'model' to consist of my chart and parallel rulers and the lines I draw. But at a deeper level my model is a mini-theory incorporating (at least) some geography, assumptions about the reliability of my chart, and geometric principles underwriting the use of the parallel rulers.

course I plot will bring me within a certain distance of my destination. If the costs of missing it are high, then I will want this distance to be relatively small and my degree of confidence to be relatively high. The higher the costs, the more steps I will take to build more factors into my calculations and to reduce my simplifying assumptions. I may even decide that I do not know enough about the tides and weather to risk the trip. Or I may know these well, but be unable to figure out how to calculate a sufficiently exact sequence of headings to counteract their effects. Moreover, this might prompt me to consult a mathematician rather than an oceanographer.

This example is supposed to illustrate how we combine scientific and mathematical theories to build models that we assess in the light of our previous experience with model building and the other factors we know. When we do not find that our models suffice for the applications we want to make of them, then we may try to improve upon their questionable assumptions either by making our idealizations more realistic (for example, in calculating the effects of the current I can use a tide table to tell me how it will vary at different stages of my voyage), or by improving our abilities to deal with complications (for example, instead of working with straight lines and parallel rules, I might plot my course using a navigational computer that implements a non-linear, time-dependent function of the tides and winds). This may even include starting over again with an entirely different approach. For example, I might decide to borrow a radio direction finder, use it to determine my current location every 20 minutes, and make rough and ready course corrections on the spot.

Discarding a model is not the same as falsifying the assumptions on which it is based. Before I even begin to calculate, I know that a model in which the tides exert no force during my trip is based upon a false assumption. Hence finding that such a model is not good enough is not the same as refuting its assumption about the tides. But that does not mean that we cannot use idealizations to test hypotheses. We build models based upon the hypothesis we want to test, and then decide whether they yield values fitting the data well enough, and better than models based upon competing hypotheses. Ronald Laymon has pointed out that idealizations often supply the keys to mathematically tractable applications of physical theories.[20]

[20] See his nice discussion of the starlight deflections tests of general relativity: Laymon (1984).

Transferring a refutation from a whole model to one of its hypotheses seems especially tricky, because in the face of a failed prediction, proponents of a given hypothesis *H* might fairly ask, 'Why should we reject *H* when we already know that the prediction was based upon a false idealization?' Let me use another boat example to illustrate how one might independently support an idealization while rejecting a hypothesis.

Suppose that on a calm day I motor across the Sound starting at a point where it is 10 miles wide using a boat which travels at 10 miles per hour. Let us also assume that I compute my compass heading by drawing a straight line on my chart and 'walking it' to the compass rose. If I arrive at the opposite shore 7 miles to the east of my intended destination, I will know that this cannot be due entirely to the tides, since at their strongest they run at 5 miles per hour. In other words, I will know that the bias introduced by my idealization is not enough to account for my setback, and that some of my other assumptions must be at fault. Among other things, I assumed that my chart was accurate, that I had correctly calculated my course, that none of the steel fittings on the boat were too close to the compass, and that I steered a steady course. Now, despite the fact that I have already taken the falsity of my assumption about the tides for granted, I will also have reason to doubt these other assumptions, and perhaps, ultimately to decide that one or more were false. Thus 'good sense' can lead one to reject a hypothesis on the basis of a failed model that contains other idealizations (false assumptions) that one knowingly retains.[21]

With this holistic model of theory testing at hand, let us now return to the objections to Quine's extension of holism to mathematics.

[21] What about scientific explanation? When Newton derived Kepler's law that the orbit of a planet moving around the Sun is elliptical, he did so by ignoring the gravitational effects of the other planets and heavenly bodies. What made this 'good enough' even though it was based upon false assumptions? I am not sure how to answer this. From the perspective of the sailing example, explanations seem to be a bonus that comes with a theory that makes good predictions. When we want to avoid or bring about repetitions of an event, such as an aeroplane crash, it is important for us to have a detailed and accurate explanation of the event. This will favour explanations based upon theories that are good predictors. From this perspective two things favoured Newton's explanation. First, it was based upon a theory that was useful in predicting; second, although he knew that the orbits could not be exact ellipses (Newton's theory told him that), the model fitted his data as well as any other contender.

Sober is right that in practice we rarely, if ever, put mathematical laws to the sorts of specific tests that we apply to some scientific hypotheses. But this does not imply that purely logical considerations show that mathematics is immune to such testing. Because entire models, rather than individual hypotheses, bear logical and statistical relationships to experience, in order to conclude that an experiment bears upon a particular hypothesis of a model we must take its other assumptions, including its idealizations, for granted. Typically justifying this will bring in pragmatic considerations, such as theoretical simplicity and mathematical tractability. Furthermore, the success of scientific model building does support our *practice* of using mathematics to formulate and develop those models. Moreover, if, as I argued in Chapter 3, this practice commits us to the truth of the mathematics it deploys, its success also supports our acceptance of this mathematics. Thus it is on pragmatic grounds rather than logical ones that we shield mathematics from revision when it occurs within a failed model.

Still, holists must acknowledge that Sober, as well as Chihara, Maddy, and Parsons, are right: it is difficult to see the success of such models, whether it be predictive, explanatory, or technological success, as providing evidence, in any ordinary sense, for the individual mathematical principles used in scientific models or as providing anything that mathematicians would recognize as mathematical evidence. From a practical point of view, we may be justified in believing that the mathematics used in science is true, but we still need an account that will reconcile this global pragmatism with our usual methods for supporting mathematical claims. I will provide the foundations for such an account in the next section and develop it further in the next chapter.

4. GLOBAL AND LOCAL THEORIES

What we take for granted in applying our models varies with the context. In sketching my last example I presumed that it was reasonable to take it for granted that the tidal currents in Long Island Sound never exceed 5 miles per hour. But what I there assumed never happens probably could happen, and it might even be reasonable to speculate that it did happen in trying to explain some cataclysmic event.

In practice, scientists working within a given context take large blocks of theory for granted. Specialized scientific theories, for example, molecular biology, are developed within a framework which draws upon principles of more general theories, such as chemistry, physics, and mathematics. Corresponding to this practice we may roughly rank the sciences in terms of their scopes and methodologies as more or less global (or correlatively, as more or less local). Mathematics is our most global theory as it is presupposed by physics, which in turn is presupposed by chemistry, etc.[22] We not only use mathematical truths in physical derivations, we also use mathematical standards to criticize physical arguments and theories (for example, we complain about the mathematical respectability of quantum field theory).[23]

Along with this rough division of the sciences a division of labour has evolved: mathematicians normally do not meddle in physics nor physicists in mathematics, and biologists and chemists are normally not competent to suggest changes in mathematics or physics even when they might want to see them changed. As a result when something goes awry in a relatively local science (say, biology), it is not likely that practitioners of more global sciences (say, physics or mathematics) will hear of it, much less be moved to seek a solution through modifying their own more global theories. Nor is it likely that the specialists in a local theory will tinker with global background theories to resolve local anomalies.

This is not just a matter of sociology, it is good sense too. *Practical rationality* counsels specialists to attempt to modify more global theories only as a last resort; for they probably do not and cannot know enough to tackle the task, and modifying a more global theory is likely to send reverberations into currently quiescent

[22] I don't count logic as our most global theory because I don't take the set of logical truths as constituting a theory. On the other hand, logical theory is metalogic, which, as it is practised these days, is a branch of mathematics.

[23] I realize that this ranking is not entirely realistic. On the one hand, mathematicians' complaints about physicists' definitions or arguments frequently fall upon deaf ears; thus physicists do not always adhere to the more global standards of mathematics. On the other hand, physicists frequently use instruments, such as cloud chambers, whose evidential status presupposes more local theories, for example, the chemistry of gases.

Despite the term 'more global', the ranking is only a partial ordering. As currently practised, neither political science nor molecular biology presuppose each other—I presume.

areas of science. Quine has expressed the point by saying that in revising their theories, scientists should minimize mutilation.

Specialization has also fostered local methodologies and standards of evidence. These provisionally override more global and holistic perspectives and declare data, obtained via local methods, to bear on this or that local hypothesis. I should emphasize, however, that on the account I am proposing, local conceptions of evidence, in particular those that lead us to take data as confirming specific hypotheses, are ultimately justified pragmatically in terms of their ability to promote science as a whole and not on some a priori basis. Hence the divisions we find in the practice and scope of the various sciences should not be taken as refuting holism or as indicating hard and fast epistemic divisions between mathematics and the so-called empirical sciences. Nor do they show that it is invariably irrational to modify some global principle to fix a more local problem.

Holists applaud the practice of physicists taking mathematics as (provisionally) fixed. But instead of grounding their approval on the supposed apriority of mathematics, they ground it on its more global status and the deleterious effects on science of revising it.

The methodological picture I have been sketching also allows for a kind of localized holism, that is, one in which scientists look at their speciality as a whole without taking into consideration how accepting a hypothesis or positing new objects or processes might influence other sciences, whether they be more global or more local. For example, although the conversion from Newtonian mechanics to quantum mechanics eventually reverberated in chemistry, physical considerations and evidence rather than chemical ones motivated the change. Similarly, purely mathematical considerations—such as the need for solutions to equations—led to many of the great advances in mathematics—such as the introduction of the complex numbers—rather than their eventual and unforeseen, widespread scientific applications.

In allowing that scientists may take a perspective that is locally holistic, I do not want to suggest that this is invariably the best thing for them to do. The early debates about the axiom of choice and non-constructive mathematics, for example, focused on philosophical and methodological issues restricted to mathematics. But we now know that rejecting the axiom of choice or non-constructive mathematics would have forced very significant changes in mathematical physics too.

When scientists posit new systems of objects and laws governing them, they also provide for connections (bridge principles) with systems of statements whose truth they already accept with a reasonable degree of confidence. Doing so enlarges the conception of evidence pertaining to the new objects, and permits them to raise and answer questions that previously may have made no sense. In positing new particles, for example, physicists will ordinarily attribute them with charges, masses, and spins in part to connect their behaviour with better understood phenomena. Similarly, mathematicians try to tie new mathematical objects to old ones, for example, approximating irrational numbers by rational ones facilitates calculations with the former.

My talk of positing might remind you of Carnap's distinction between internal and external questions, so I think I should explain how my conception differs from his. According to Carnap, positing a new system of objects is really to introduce a new language along with axioms and rules of inference that enable us to decide questions formulated in that language. The language's axioms may include both physical and mathematical laws, and its rules of inference may include procedures for making empirical inferences as well as those of logical and mathematical deduction. On this picture, enlarging a local conception of evidence is tantamount to introducing a new linguistic framework. Furthermore, rejecting a framework axiom or rule, whether it be logical, mathematical, or 'empirical', amounts to rejecting the framework as a whole. He also believed that questions about the acceptability of a framework—he called them 'external questions'—are fundamentally different from those raised within a framework—his internal questions. The rules of a framework fix the considerations relevant to deciding its internal questions, and nothing else counts, whereas only pragmatic considerations are pertinent to deciding whether to accept a framework itself.[24]

Carnapian frameworks are built from scratch, so that we do not just tack, say, quark theory, onto a pre-existing quantum mechanics, but start out by specifying the logical and mathematical principles of quark theory, and then its fundamental physical laws, and so on, until we introduce the specific quark hypotheses. Of course,

[24] Actually, Carnap replaces the distinction between the logico-mathematical and the empirical with a framework-relative analytic–synthetic distinction. The analytic truths are those that follow from the postulates constituting the framework via its logical rules. These rules are in turn introduced by stipulation. See Carnap (1956).

'hypotheses' is a misnomer, since in the quark framework the quark hypotheses are just as a priori (and analytic) as its logical laws. Since Carnap recognized no facts prior to or transcending frameworks, framework principles cannot be true in virtue of independently obtaining facts. Instead, they are true by convention.

Unlike Carnap, I see global theories as transcending the local theories that are build upon them, and as provisionally fixing a background of facts upon which local theorists may draw. These facts may also furnish them with reasons for preferring one local theory to another. They might show, for example, that a proposed theory is untestable.

Carnap did not seem to consider the possibility of appealing to pragmatic grounds to justify adding a new hypothesis to an extant framework, as we might do today in positing, for example, a new type of quark, or a larger infinite cardinality. This is just where Quine would argue that there is no real epistemic difference between 'externally' changing a framework by adding a new hypothesis (expressed in the language of the framework) and 'internally' deciding that the evidence supports the hypothesis.[25] I side with Quine on this.

Quine sometimes gives the impression that the decision to accept a new hypothesis is always to be referred to the system of science as a whole. I can agree to this too, but only if we make the extra methodological layers explicit. The decision to accept a hypothesis is often dictated by local methodological principles that in turn have developed from a more global perspective, and that perspective in turn may involve an even more global one, and so on, until we ultimately reach considerations affecting the system of science as a whole.

Let me now review some of our earlier methodological observations using the global/local distinction. Holism should be uncontroversial if construed as a comment on the logical relationship between theoretical hypotheses and their empirical consequences. To derive a prediction *P* from such a hypothesis *H* we must not only appeal to *H* itself but also to auxiliary assumptions *A*. If *P* fails, all that follows is that *H* or *A* is false. To conclude via disjunctive syllogism that *H* is false, we must already take *A* to be true.

Yet it is certainly not practical for scientists to treat every hypoth-

[25] See Quine (1962).

esis as equally vulnerable each time they run an experiment. The path to success lies in taking a more local perspective and tinkering with those hypotheses one understands best. From such a perspective it makes sense to treat more global auxiliaries as (provisional) fixed points. Thus, for example, when some biologists use some mathematics and a biological hypothesis to design an experiment which produces results contrary to their expectations, there will be (and should be) no question in the mind of the biologist but that it is the biological hypothesis which is at fault. This is no refutation of holism, but merely an illustration of the global/local approach to methodology.

I can now say something about the relationship between holism and our ordinary conception of mathematical evidence. Empirical success no more confirms individual mathematical claims than it does individual theoretical hypotheses. However, it does provide a pragmatic justification for positing mathematical objects, truths about them, and principles for applying mathematical laws to experience. From this perspective we may encourage mathematicians to develop their own standards of evidence, so long as the result does not harm science as a whole. Because mathematics is our most global science we should expect that many mathematical methods and principles will be justified by means of considerations neutral between the special sciences, and thus often pertaining to mathematics alone. In this way we can reconcile holism with the features of mathematical practice that Chihara, Maddy, Parsons, and Sober have emphasized.

Considering the place of proof in mathematics will illustrate this. Early mathematicians probably took their experience with counting, book-keeping, carpentry, and surveying as evidence for the rules and principles of arithmetic and geometry that they eventually took as unquestionably true. They began to put more emphasis on deduction after they became aware of the difficulties in deciding certain mathematical questions by appealing to concrete models, which, for example, are notoriously unreliable in deciding geometric questions. (How could we know this without independent access to geometric truths? By persistent disagreements among the answers the models supply, or by their inability to supply answers at all.) Even today we could (and sometimes do) use concrete models to decide certain mathematical questions; for example, we might simulate a Turing machine on a computer to determine whether it gets into a certain state when processing a given input. But the advantages of proof to the practice of mathematics are so obvious that frowning on experimental approaches has served the goals of mathematics better than allowing or promoting such approaches. Moreover, proof wins out from the perspective of science as a whole. For requiring mathematics to prove its results increases its reliability, and decreases its susceptibility to experimental refutation.

The development of non-Euclidean geometries and abstract algebra further promoted the purely deductive methodology of the axiomatic method through showing mathematicians how to make sense of structures that might not be realized physically. It also promoted a shift from viewing mathematical sentences as unqualifiedly true or false to regarding them as true or false of structures of various types. These two developments have further insulated mathematics against empirical refutation. To see how, consider the case of Euclidean geometry. General Relativity *did* refute it in its original role as a theory of physical space, but it still has important mathematical models, and survives through being reinterpreted as a theory of Euclidean spaces. (With hindsight we say that we discovered that physical space is not Euclidean.) Now a similar move is available when a scientific model incorporating a bit of mathematics proves inadequate to a physical application. It is usually far simpler to save the mathematics from refutation and conclude that the physical situation to which it was being applied failed to exhibit a suitable structure. (For obvious reasons, I will subsequently refer to an instance of this way of saving a theory as a *Euclidean rescue*.) We can apply a Euclidean rescue to any theory by reinterpreting it, which is what we did for Euclid's geometry. Mathematical theories need no reinterpretation, however, since they do not assert that the structures they describe are realized in this world. Ironically, the ease with which Euclidean rescues may be applied to mathematical theories tends to encourage those separatists who regard mathematics as immune from empirical refutation.

5. REVISING LOGIC AND MATHEMATICS

Holists and separatists subscribe to different doctrines concerning the nature of mathematical and scientific evidence. Despite this, we are unlikely to find that differences in the attitudes of most holists towards individual scientific decisions or the practice of contemporary science and mathematics suffice to distinguish them from separatists. Thus, for example, both are likely to recommend using a Euclidean rescue to save the mathematics of a failed scientific model. This is what one would expect given that holists can accommodate separatist intuitions concerning mathematical practice. Since contemporary objections to holism have been based upon these intuitions and not upon a developed mathematical epistemology, showing that holism and separatism agree on actual cases significantly weakens the case for the latter.

By introducing some new characters, the strict holists, we can bring out a difference between holist and separatist views concerning specific scientific practices. Strict holists maintain that we have no evidence for a claim unless it is part of a system of hypotheses from which confirming observations have been deduced. They do allow for local conceptions of evidence. But they prohibit those conceptions which allow evidence to accrue to specific hypotheses that are not part of a confirmed system of hypotheses. In particular, they prohibit the introduction of other (for instance, non-observational) evidence for scientific or mathematical claims. On this view, mathematics with no current application, such as results employing large infinite cardinals, would have little title to truth.

One problem with strict holism is that it is not clear that it could countenance certain metamathematical investigations of even the mathematics that does get used directly in science. It is hard to see how investigations of the consistency of various systems of analysis could be directly useful in *constructing* scientific models or in *deriving* empirical consequences from physical hypotheses formulated in the vocabulary of analysis. Despite this, the information these proofs provide about the reliability of those models is relevant to *evaluating* the sciences using them. Similarly, it is often useful for scientists to know of mathematically equivalent ways of formulating scientific theories or of probabilistic approximation methods that might substitute for rigorous calculations. Yet in deriving empirical results they can often use alternative formulations directly without having to cite metamathematical equivalence proofs. This is one reason why I would advocate a liberal version of holism, one which would allow the development of a local conception of mathematical evidence that could countenance mathematical truths that have no foreseeable empirical use.

Some of Quine's remarks on higher set theory might suggest that he is a strict holist. But he is not. Consider this passage:

What now of those parts of mathematics that share no empirical meaning, because of never getting applied in natural science? What of the higher reaches of set theory? We see them as meaningful because they are couched in the same grammar and vocabulary that generate the applied parts of mathematics . . . On our two-valued approach they then qualify as true or false, albeit inscrutably.

They are not wholly inscrutable. The main axioms of set theory are generalities operative already in the applicable part of the domain. Further sentences . . . *can still be submitted to considerations of simplicity, economy, and naturalness* that contribute to the molding of scientific theories generally. Such considerations support Gödel's axiom of constructibility, 'V = L'. It inactivates the more gratuitous flights of set theory, and incidentally it implies the axiom of choice and the continuum hypothesis.[26]

Although the first paragraph clearly shows that Quine believes that securing evidence for unapplied mathematics is problematic, in the second paragraph he goes beyond the strict holist in allowing pragmatic considerations to justify accepting axioms that are not needed in deriving empirical consequences from scientific hypotheses. In particular, if ZFC alone suffices for the mathematical needs of natural science, then from the strict holist's point of view there is no further need to decide the continuum hypothesis or to inactivate 'the more gratuitous flights of set theory'. We can get along with ZFC alone without adding any further axioms to address the issues Quine thinks we should.

What other differences might there be between holists and separatists? Strict holists can conceive that a mathematical statement we previously regarded as true might degrade to one whose truth-value cannot be settled, because it comes to fall into an area of mathematics no longer connected to experience. Thus, according to them, it may not always be possible to perform a Euclidean rescue. Quine, though no strict holist, envisages something like this in remarking that the mathematics we recognize as true might be cut back to predicative set theory if it can meet the mathematical needs of science.[27] Separatists, of course, would not permit such developments to curtail otherwise acceptable branches of mathematics.

[26] Quine (1990), 94–5 (my italics). [27] Ibid.

I am merely speculating, to be sure. I know of no strict holists nor of any mathematics that science clearly no longer needs. But it is to such speculations that we must turn to discern other differences between the recommendations of holists and separatists. Of course, the doctrinal differences between them are obvious.

Holists can allow an experiment to refute the mathematics used in designing it by waiving local conceptions of evidence. How it might refute the mathematics in question would depend upon the case at hand. Let us consider some of these cases.

First, let us note that it can happen that scientists use a mathematics so rudimentary and tentative that both its concepts and proof methods fail to be well defined. This is a reasonable view of the calculus used in late seventeenth- and early eighteenth-century physics. Now if we developed a new branch of mathematics for a specific type of application and using it proved unreliable, then rather than performing a Euclidean rescue it might be more rational to reject the mathematics as ill-formed. Since this option is open to both holists and separatists, I will henceforth restrict my attention to rigorously specified mathematical theories.

Suppose that while designing an experiment some physicists need to know whether a certain function has a maximum. They hear about a conjectured consequence S of a new extension of ZFC set theory, ZFC + A, using a new axiom A, and they subsequently determine that S implies that their function does have a maximum. Let us also suppose that they assume this maximum exists in testing one of their pet physical hypotheses and that their experiment produces a result conflicting with their expectations. We can imagine them deciding that the conjecture must have been wrong—not simply that they had misunderstood or misheard the conjecture.

But what does it mean to say that the conjecture is wrong? If one can be reasonably certain that the set theory in question (ZFC + A) applies to the physics in question, then it might be reasonable to take the experiment as evidence that the conjecture is not a theorem. Now a separatist might hold that certain empirical data, such as computer outputs or the results of hand computations, could refute or establish a conjecture.[28] But holists must allow that even experiments, such as those used in physics, that were not originally

[28] These separatists might deny that mathematics is a priori but still claim that only certain empirical data count as mathematical evidence.

designed to test a mathematical conjecture might count against it. For they countenance no a priori evidential distinction between experiments that are designed to test a specific claim and those that merely presuppose the claim without testing it.

Perhaps the physicists will conclude that the conjecture was wrong in a weaker sense of being inappropriate to their physics. Then they might conclude that the set theory in question does not apply to the physics—that it does not exhibit the appropriate set-theoretic structure—and they will execute a Euclidean rescue.

Let us now suppose that instead of a conjecture the problematic mathematical claim S is a known theorem of the extended set theory, ZFC + A. Then the physicists would almost certainly perform a Euclidean rescue. For what other options would they have? They might examine the theorem's proof on the off-chance that it contains an error; but, due to their relative lack of mathematical expertise, they would have little reason to expect this tactic to succeed.

Finally, let us imagine that after repeated unsuccessful efforts to account for the anomalous experiment our physicists call in the mathematicians. Even then it would be rational for the mathematicians to try a Euclidean rescue first, and to attempt to apply to the physics a different, perhaps newly developed branch of mathematics.

Would it be rational for the mathematicians to see the experiment as refuting ZFC + A? According to our story, it implies a claim that when combined with physical hypotheses forms a package contrary to the experimental results. Stretching our imaginations to near the breaking-point, we can think of the mathematicians as arguing as follows: some statement in the package must be false, and the physical ones are beyond question, so the mathematical statement must be false; but since this is implied by ZFC + A it must be false too. But notice how strong a claim this would be. Because one can always save a consistent branch of mathematics via a Euclidean rescue (and we have assumed that our mathematicians have excluded this), for them to reject the axioms of ZFC + A would be to take them to be inconsistent![29] Of course, they could avoid this step by taking an

[29] In drawing this conclusion our mathematicians had assumed that ZFC + A applied to the physics. Instead of concluding that it is inconsistent, shouldn't they just conclude that it does not apply after all? Yes, this would be a more reasonable choice if they had independent grounds for thinking that ZFC + A was consistent. But holists could find the stronger conclusion warranted if the mathematicians' independent grounds for believing in the consistency of ZFC + A were weak.

even more radical one. They could take the physical model as showing that the axioms do not imply the theorem, and then proceed to modify the rules of deduction used in its proof.

The main difference between holists and separatists arises when a well-established piece of axiomatic mathematics is part of a model conflicting with experience. Separatists recognize only the Euclidean rescue, while holists also admit the options of rejecting the mathematics as inconsistent or altering its underlying logic. They see nothing in the epistemology of mathematics that excludes using these options in the rare instances when doing so will benefit science more than the Euclidean option.[30]

Of course, the foregoing does not refute separatism; it only helps make holism more palatable to those with separatist leanings. Moreover, even this tempered and conservative holism would be unacceptable if, independently of holism, one could establish an epistemically principled division between the empirical and formal sciences. But I do not see much hope of success here. Consider the difficulties confronting separatists.

First, they would need a way of dividing mathematics from the rest of science which explains why we should not use Euclidean rescues to save sufficiently precise scientific theories. Otherwise, it would be reasonable to hold, for example, that the Michelson-Morley experiment merely showed that bodies travelling close to the speed of light do not form Newtonian systems. I like this move for the same reason I like using Euclidean rescues in mathematics: it preserves more truths and more theories.[31]

[30] This is not to deny that separatists can hold that unexpected experimental results might prompt us to re-examine our mathematics or logic with the possibility of revising one or the other. But separatists are committed to holding that the grounds for revising either would be non-empirical ones discovered as a result of re-examining them rather than the experimental results prompting the re-examination.

[31] In correspondence, Mark Balaguer has suggested that blurring the epistemic boundaries between mathematics and empirical science depends upon successfully blurring their ontologies. For suppose that all the objects studied by science are causally active denizens of space-time, but none studied by mathematics are. Then the mathematical realm has no effect on the observable realm and conversely, and so observational data have no bearing on the truth of various mathematical claims.

Now this conclusion is correct if it simply means that events in the empirical world have no effect on the truth-values of mathematical claims. But it does not follow from this that observation is irrelevant to determining their truth-values. For if a system of physical objects realizes a given mathematical structure, then examining structural features of the system can inform us about the mathematical structure.

Second, they would also require a similar division between logic and mathematics, or if they lump mathematics with logic, a division between logic and the rest of science. A stronger logic sees more distinctions between structures, decides more mathematical propositions, and attributes more empirical consequences to a theory. A stronger logic thus narrows our options for revising theories. Since a weaker logic requires us to supplement a theory with more auxiliary hypotheses to test it, it leaves us with more options when our predictions fail.

Third, the division between the formal sciences, logic and mathematics, and the rest of science must be more than just a matter of methodological convenience for separatists; otherwise they will be indistinguishable from holists. To date, there is no uncontroversial principled division between the formal sciences and the empirical ones.

In the next chapter I will argue that when it comes to logic, there is no such division because there are no special logical facts to be known. The truths of logic are just ordinary truths generated via our most global methodology, and in calling something a logical truth we do not ascribe a metaphysical property to it, but rather mark it for special treatment.

8

The Local Conception of Mathematical Evidence: Proof, Computation, and Logic

In the last chapter I noted elements of our scientific practice supporting separatist theses: we often take our experiments to test specific statements instead of the larger theories to which they belong; we usually agree that a piece of empirical evidence pertains to one claim and not to another; and we generally allow mathematics to remain aloof from the empirical fray, shielding it from empirical disconfirmation and testing. I argued that we can explain this practice within the framework of pragmatic holism by pointing out that the local conceptions of evidence, which underwrite these practices, are necessary for the development and progress of science as a whole. Thus these practices can be seen to derive from considerations of practical rationality rather than from those of logic or some other a priori source. Consequently, the facts of scientific practice by themselves do not entail that mathematical knowledge is fundamentally different from so-called empirical knowledge.

At the beginning of Chapter 6 I raised several questions concerning mathematical evidence which I have yet to answer. What kind of evidence do we recognize for our mathematical beliefs? What roles do logical deduction, proof, and computation play in justifying these beliefs? How do we justify introducing new mathematical axioms or new kinds of mathematical objects? Although I said earlier that my answers to questions such as these would presuppose holism, we can see now that the situation is more complicated than I had indicated. For these questions concern the local conception of mathematical evidence, the conception governing ordinary mathematical practice. Ordinarily, mathematics is practised as an autonomous science, and its evidential norms do not refer to science as a whole. On the face of it, this fits poorly with my holistic account.

For this reason I will begin this chapter by showing in more detail than I have so far that the features of mathematical practice to which separatists appeal can be brought under the purview of a holistic account of mathematics.

I will also argue that, holism aside, mathematical practice is not free of either non-deductive or empirical methods, and that some empirical content can be given to certain mathematical claims. Finally, I will contend that the division between logic and the rest of science and mathematics is conventional, and that the dividing line and individual logical principles are revisable in principle. Thus even purely logical deduction of a theorem from a set of axioms does not yield a type of knowledge that is immune to revision.

1. SOME NORMS OF MATHEMATICAL PRACTICE

What kind of evidence does our local conception of mathematical evidence recognize? Certainly, at least proof and computation. In practice most 'proofs' are at best sketches of the steps required to deduce their theorems from an appropriate set of axioms, but none can pass muster if such a deduction is in doubt. Furthermore, associated with every computation is a deduction from theorems authorizing the steps of the computation in question. Thus in theory, if not in practice, both proofs and computations can be considered as series of deductions from axioms.[1]

But clearly mathematics is not special in accepting proof, computation, and deduction as evidence, for other sciences extensively employ these methods too. What is special about mathematics is that so long as it seems possible to prove (or refute) a conjecture, mathematicians regard it as an open question, even when, by the standards of the natural sciences, the non-deductive evidence deciding the result one way or another is overwhelming. If, for example, decades of computer runs regularly produced new pairs of twin primes, it is unlikely that mathematicians would regard the twin prime conjecture as established. Yet natural scientists would surely regard similar experiments as decisive.[2]

[1] For a useful discussion of the relation between formalization and mathematical practice see Steiner (1975).

[2] This type of case differs interestingly from the non-deductive evidence favour-

To be sure, mathematicians might be convinced by such evidence, while still taking the question to be officially open. I should add such that Lakatos, Putnam, and Steiner have described historical examples purporting to show that mathematicians are willing to decide questions on the basis of non-deductive inference.[3] Moreover, some contemporary mathematicians speculate that increased computer testing will lead to the acceptance of non-deductive substitutes for proofs.[4] Finally, certain claims about the behaviour of computer programs are mathematical claims, and computer runs are sometimes regarded as decisive evidence concerning them. I will postpone discussing cases of this sort until the next section.

Now it is not enough for separatists to point out that mathematicians insist on proof where other scientists do not, because pragmatic holists can argue that the practice in mathematics makes 'good sense'. For emphasizing proof promotes systematization and reliability in both mathematics itself and those branches of science applying it.

Although scientists and mathematicians use deduction, the former use it (largely) to prove theorems from axioms while the latter use it (largely) to derive testable consequences from hypotheses. Separatists might try to argue that this difference is due to the differing character of the premisses used in science and mathematics. It might be that while the evidence supporting the theoretical assumptions used in science is tentative, that supporting our mathematical axioms is conclusive. However, if the opinions of mathematicians are any indication, some axioms are more securely established than others. For example, hardly anyone seriously objects to the axioms of elementary number theory, but many mathematicians regard axioms postulating large infinite cardinals as speculative, and a few others suspect even impredicative analysis.

Furthermore, there is no clear understanding among mathematicians as to what counts as conclusive evidence for a mathematical axiom. Many philosophers and philosophically inclined mathematicians have written of axioms as self-evident, obvious, or intuitively

ing the Riemann hypothesis, which computation cannot directly verify. For further discussion of the evidence for the Reimann hypothesis see Davis and Hersh (1981).

[3] See Lakatos (1978), Putnam (1975), and Steiner (1975).

[4] See Horgan (1993).

known. But this helps little, if only because mathematical researchers differ radically in assessing these properties. (And these notions are in desperate need of clarification themselves.) In addition, mathematicians sometimes argue for admittedly 'unintuitive' axioms on other grounds, such as their ability to systematize large collections of results or to yield theorems in more secure areas of mathematics.[5]

What is more, it is sometimes not clear whether the evidence mathematicians cite for axioms concerns their consistency or their truth. Penelope Maddy presents the debates among set theorists as if they concerned the truth of various axiomatic extensions of set theory,[6] and I do not dispute that this is how the participants regarded them. Yet one can also interpret the discussions she reports as really concerning a choice between consistent ways of extending set theory. Instead of seeing set theorists as seeking the true set theory, we might interpret them as seeking the most desirable one, where, of course, we understand desirability in terms of mathematical fruitfulness, elegance, simplicity, and the other virtues, falling short of truth, that set theorists cite in favour of their proposals.[7] And if one wants to bring in truth here, one can let it concern what is *true in* that alternative set theory that best settles the supposed indeterminacies in our current conception of set.[8]

So far we have found little in mathematical practice to clarify the nature of the evidence, if any, for mathematical axioms. It will help to consider the matter in the light of the types of problems mathematicians tackle. Most of these problems focus on questions arising within the confines of familiar mathematical theories, and concern features of the mathematical domain or domains these theories cover. Answering these questions consists in proving theorems within axiomatic systems. Here the truth of the axioms is not a live issue. The real issue concerns what is true if they are true, and in the course of proving theorems one provides conclusive evidence for such conditional truths. Of course, if one already believes the

[5] See Maddy (1988) for an excellent survey of the considerations set theorists have introduced in discussing proposed extensions of standard set theory.

[6] Ibid.

[7] For a critique of Maddy's interpretation see Riskin (1994).

[8] Ironically, some mathematicians appeal to holist considerations in condemning the 'excesses' and 'theology' of transfinite set theory. For they lament that contemporary mathematics has lost its roots in applications, lacks direction, is sterile.

axioms, it is easy to confuse these proofs with conclusive evidence for the theorems themselves.

Another type of problem calls for using axioms to characterize a concept or structure (such as the concept of a group, the natural number sequence) or a body of results or techniques (such as predicative analysis, school arithmetic) that are already on hand. Here we want evidence that the axioms are the 'right' ones. A necessary condition for some axioms to be right is that they yield theorems in the body of results in question (or theorems that are acknowledged features of the structure being characterized). So once again proving theorems, using the axioms in question, provides an important part of the evidence that one has a good answer. Furthermore, if it is clear that the axioms themselves belong to the results in question (or formulate acknowledged features of the relevant structure), one might feel confident that they are among the 'right' ones. But this usually does not suffice to solve the problem. We usually want to know whether the axioms produce all and only the appropriate theorems, and if they do not, then why not. Or we may want to know whether they are categorical. Furthermore, what counts as having the 'right' axioms varies with one's other goals. Dedekind's categorical axiomatization of arithmetic provided no solution to the problem Frege set himself, because he did not believe that the Dedekind axioms were purely logical. (Their primitives include *zero* and *successor*.) In problems like Frege's, then, proofs within an axiom system usually will not suffice as conclusive evidence for a solution. Also we may be unable to formulate constraints on our axioms with sufficient precision to prove metatheoretically that they are the 'right' ones.

A third type of problem arises when mathematicians set up and explore new axiom systems in order to introduce new concepts and structures or to modify old ones. In this case they hope to demonstrate that the new axioms are worth exploring—that they are consistent, have the 'right' sort of models, bear 'interesting' relations to old systems or structures, and so on. This requires them to take a more openly metamathematical perspective than they might when dealing with problems of the first or second type. But even here they prove theorems (and metatheorems) to provide evidence concerning various properties of their axioms.

One could construe even these metatheoretic investigations as cases of proving theorems within some axiomatic metatheory. On this view, investigations of the second and third type reduce to the first. The only evidence mathematicians would need is that something is a theorem of the appropriate system, and proof would be conclusive. I have called this position *deductivism*;[9] others call it 'if, then-ism'. By whatever name it goes, it is an unsatisfactory doctrine. Mathematicians want to know that their systems have models; and they want to know this absolutely, and not just relative to a metatheory. Furthermore, as we saw in Chapter 3, applying mathematics to science presupposes that mathematical models exist—again absolutely and not merely relative to a set of metatheoretic assumptions. Thus the methodology of mathematics and science takes a realist stance towards at least some mathematical theories.

We are left committed to a plethora of mathematical theories, many of which are incompatible with one another—as are, for example, the alternative geometries and set theories. We can resolve the apparent inconsistency in our commitments by restricting these theories to structures of the appropriate type—Euclidean geometry to Euclidean spaces, ZF set theories to iterative hierarchies, and so on—so that each theory is construed as true of only the appropriate structures. But, at least from the methodological point of view, mathematics is committed to a plurality of structures and to the appropriate theories' being true in the appropriate structures.[10] What evidence, then, do we have for these structures and truths concerning them?

Given some structures we can prove that others exist by constructing models of them within the former. This technique won acceptance for the imaginary numbers and non-Euclidean geometries, and established the relative consistency of the axiom of choice. But, of course, this kind of support is no better than the evidence we have for the systems within which the constructions take place. This is why we try to use assumptions in which we have the most confidence, such as those used in the well-entrenched portions of number theory, geometry, analysis, and set theory. Even here there is reason to be more confident in some methods, models, or theories than others. Simple finite models, for example, are easier to understand and check than infinite ones based upon sophisticated mathematical

[9] See Resnik (1980).

[10] I say 'at least from the methodological point of view' since one might give a philosophical account of mathematics which shows that these assumptions needn't be taken literally.

constructions. (Compare these: the consistency proof for the predicate calculus, Gentzen's consistency proof for number theory, Gödel's consistency proof for the axiom of choice, Cohen's independence proof for the same.)

Now why should we be more confident in elementary arithmetic and geometry than in the theory of infinite cardinals? One reason is that we can often compute some instances to convince ourselves of arithmetic formulas or draw diagrams to illustrate geometric theorems. We can also test rules for computing by counting or manipulating collections of physical objects, and geometric theorems by measuring. While official mathematics disdains evidence of this sort, it certainly plays a more crucial role in convincing beginning students of mathematics than proofs do, and mature mathematicians are not above using it when they can. What is more, this sort of verification is supposed to be the historical evidence for geometry and arithmetic.

Another reason for being more confident in our older, more entrenched mathematical theories is that they have been much more extensively applied than the newer theories, which usually not only are logically stronger but also have no known or intended applications. So if the older theories were based upon ill-defined notions or inconsistent assumptions, the chances are good that trouble *attributable to them* would have arisen within the scientific theories that use them.[11] While this is no 'direct' test of this mathematics, it the same sort of indirect test that our more general scientific theories receive in being used in constructing more specific theories or models.

The picture I have been sketching is one in which the more theoretical branches of mathematics draw their support from the more elementary and established branches, which in turn draw their support from their success in applications and, historically, from empirical demonstrations aimed at supporting certain elementary claims. If this is correct, then *even from the local point of view* some of the evidence for the most elementary and fundamental parts of mathematics is empirical. The difference between the natural sciences and mathematics is not that the former recognizes evidence which the

[11] For contemporary examples of mathematics faulted in its applications consider the many special-purpose algorithms programmed into computers, whose use often reveals inconsistent instructions, false presuppositions, and neglected cases.

latter prohibits, but rather that this evidence no longer plays the role in contemporary mathematics that it played historically. The evidential roots of the older branches of mathematics are empirical, but these branches now count as evidence themselves for the newer, more speculative branches.[12]

Have we not returned to the point Sober raised against holism? That only those tests specifically designed to test a hypothesis should count for or against it? In a way, yes, but here I am arguing that even those who separate mathematics from science should recognize that Sober's point is false. Let me liken hypotheses to tools. Tool manufacturers are likely to put their products to specific laboratory tests, for instance, to determine how much force is required to break them. This is like laboratory testing, say, the accuracy of the inverse square law for charged particles. Tool manufacturers conduct such laboratory tests to determine whether to market their tools and how strongly to warrant them. But a tool's performance in service also counts as evidence for its reliability and limitations, and manufacturers modify their warranties and operating manuals in the light of such evidence. By analogy, using a hypothesis (even as an auxiliary) to construct successful models tends to support the hypothesis just as successful service supports the reputation of a tool. Moreover, just as using a tool in a failed project can still provide some favourable experience with the tool so long as there is no obvious reason to blame it for the project's failure, so too can failed models in which there is no obvious reason to blame a hypothesis count a bit in favour of it.[13]

No one disputes that by using appropriate auxiliary premisses we can derive empirical consequences from certain mathematical hypotheses—that, for example, we can test claims about Turing machines by simulating them on electronic computers. In so far as

[12] In correspondence Mark Balaguer has questioned the use of applications to justify mathematical theories on the grounds that these theories are justified prior to applications. Of course, there are famous examples of mathematical theories—group theory is one—which were well justified prior to any applications. But I think it is equally true that arithmetic, geometry, and even the calculus devloped along with their applications. The successful applications of the calculus probably helped set aside doubts about its foundations. In describing the evidential roots of the older branches of mathematics as empirical I had in mind examples of this sort.

[13] To push the analogy further, notice that wringing the head off a bolt does not count against the wrench, but rounding the head can count against both the bolt and the wrench. So one can imagine cases in which failure counts against several of the hypotheses.

separatists count similar tests of scientific hypotheses as directly testing them, such a test should count as directly testing claims about Turing machines. Thus we can meet separatists on their own ground and design experiments to test certain specific mathematical hypotheses. Of course, we cannot design experiments to test every mathematical claim, but there are plenty of scientific hypothesis that we cannot test either. For example, there is no way to determine whether physical space is a continuous manifold.

Separatists might still object that mathematics has no empirical content of its own despite examples such as these. They could point out that empirical science aims to describe and explain the world about us while mathematics, even when playing its indispensable part in furthering this aim of science, aims to describe all possible worlds. We can put this point in more precise terms by supposing that we can divide our vocabulary between purely physical terms ('rabbit', 'apple', 'yellow', 'metal') and purely mathematical ones ('number', 'derivative', 'average'). Such a division would determine three classes of statements: the purely physical statements, the purely mathematical statements, and the mixed statements. Purely mathematical statements will have trivial mixed logical consequences. For example, 'every set belongs to a set' implies 'every set belongs to a set or gold is an element'. Conjoining some mixed truths with purely mathematical ones will yield mixed non-empirical truths. For example, conjoining 'every set belongs to some set' with the non-empirical, mixed premiss 'there is a set of planets' implies 'the set of planets belongs to some set'.[14] But no consistent body of purely mathematical statements would imply any purely physical statements except the logically true ones. Thus their failure to imply any purely physical statements would give us a sharp sense in which mathematical statements have no empirical content, whether taken collectively or individually.

I think this sort of reasoning underlies the thinking of many philosophers. But it overlooks the fact that many systems of supposed scientific statements have no empirical content on their own. One need only turn to philosophical discussions of the geometry of physical space, Newton's laws of motion, or Darwin's theory of

[14] This statement, which follows from a mixed version of the comprehension axiom, should not be confused with the empirical statement that the set of planets is non-empty.

evolution to find examples. Furthermore, it is based upon uncritically accepting the distinction between purely mathematical and physical terms. Purely mathematical statements fail to imply purely physical ones because of the separation in their vocabularies. The same is true of any two collections of statements composed of disjoint non-logical vocabularies. In view of the difficulties we saw in Chapter 6 in classing space-time points or quantum particles as clearly physical objects, the burden of providing a clear epistemic distinction between mathematics and the rest of science lies with the separatists. Plainly it will not do for them just to list the supposed mathematical vocabulary and leave it at that.

So far my account of the local conception of mathematical evidence has emphasized the evidential role of applications of mathematics. Let us turn now to the unapplied parts of mathematics.

In Chapter 3 I noted that indispensability arguments fail to cover the more theoretical and speculative branches of mathematics. In the last chapter we saw Quine arguing that, considering the mathematical needs of science, it is preferable to 'inactivate' the higher reaches of set theory by adopting an axiom, such as $V = L$, to limit the set-theoretic universe. I pointed out that indispensability considerations alone would leave the extent of the set-theoretic universe undecided. For science has no need of hypotheses deciding it.

Nor does the mathematics currently employed in science presuppose higher cardinals. However, the story is less clear when it comes to their ultimate indispensability. On the one hand, work on predicative analysis indicates that in principle contemporary natural science could get along with a weaker mathematical apparatus than it actually employs; on the other hand, other research indicates that there may be no limit to the large cardinal assumptions one might use to decide scientifically pertinent questions in real analysis.[15] At the moment indispensability considerations recommend agnosticism concerning further extensions of set theory.[16]

But we should also ask how matters stand when we turn from just indispensability considerations to the pragmatic holism cum local conceptions of evidence that I have been advocating. To begin, let us note that so long as we use set-theoretic models to study and sup-

[15] See the discussion of the mathematics needed in science in Hellman (1989).

[16] Most 'ordinary mathematics', and thus most of the mathematics used in science, does not even require the axiom of replacement. See Lavine (1994)

port the introduction of new mathematical structures, axioms limiting the size of the set-theoretic universe would discourage the development of mathematics through limiting the structures it recognizes. Furthermore, while limiting the variety of structures would probably not hinder contemporary science, it might hinder future science. So the good of neither mathematics nor science as a whole calls for adding limitative axioms to set theory.[17]

We also know that certain limitative axioms, such as $V = L$, are independent of the other axioms of set theory, and so can be consistently negated. Taking the more positive step of postulating further extensions of the set-theoretic universe, on the other hand, runs a greater risk of inconsistency. For in general the new systems are stronger than the older systems by virtue of entailing the consistency of the latter.

Still, mathematicians want to know whether these axioms can be used to enrich the class of structures mathematics countenances. (If we take the usual metatheoretical perspective and assume that mathematical structures are set-theoretically defined models, then certain sets exist just in case certain structures do.) Unfortunately, the initial evidence for these new axioms is less substantial than that for new hypotheses in the natural sciences. Natural scientists expect new hypotheses, when combined with appropriate auxiliaries, to yield previously untested observational consequences. In short, they should be capable of at least indirect empirical testing. The mathematical analogue of this methodological rule would be to require new axioms to imply new computationally testable results. But mathematics has no such requirement. And with good reason: the computationally decidable mathematical truths are completely axiomatizable; consequently, new axioms will not yield new members of this class.[18] (However, they usually imply results whose instances belong to the class, such as arithmetical consistency statements for the systems they extend.)

Instead we may find that the new axioms yield non-computational consequences which can be verified independently by using

[17] This is not to say that we cannot recognize set-theoretic structures in which limitative axioms hold. Indeed, Gödel showed that every model of ZF set theory contains a submodel in which $V = L$.

[18] By computationally decidable mathematical truths I mean the variable-free, true numerical equations of primitive recursive arithmetic plus those that can be coded as such equations.

methods that are available within older systems. These methods might be quite advanced themselves. But the more elementary they are, the more confidence we are likely to have in them, and in the indirect support their results give to the new axioms. We can also usually prove that older structures are contained within the structures the new axioms imply. Such results show that new axioms extend the old systems in appropriate and interesting ways.

Of course, we would also welcome evidence that the new axioms are consistent. Due to the strength of the new systems this will take time. Initially, the most we can expect is to show that the systems do not permit the obvious derivations of well-known contradictions such as Russell's Paradox. We might also show that the systems are consistent provided other equally powerful systems are. If we have additional evidence for the consistency of the latter systems, then it transfers to the former. Analogies to well-entrenched extensions of even older systems can also provide some initial reason to be confident in the consistency of new systems. (For example, the axiom postulating an inaccessible cardinal draws some support from its similarity to the ordinary axiom of infinity, which postulates omega.)

To summarize, if the account I have been giving is correct, then the relationship between our mathematical theories and the evidence for them is quite similar to the indirect sort of relationship holding between scientific theories in general and the evidence for them. Furthermore, the evidence for our more elementary and older mathematical theories derives—not from special mathematical insights or reasoning—but rather from the empirical consequences they (and appropriate auxiliaries) produce and their successful use in science. Thus, despite its emphasis on deductive proof, even the local conception of mathematical evidence does not differ as radically from that of empirical science as tradition holds.

2. COMPUTATION AND MATHEMATICAL EMPIRICISM

Although in the previous section I argued that some of the evidence for the old parts of mathematics is empirical, I did not claim that non-deductive inference from empirical premisses is part of the contemporary methodology of mathematics. In this section I shall appeal to the role of computation in mathematics to argue this claim. It will turn out that only an ideal being could pursue mathematics through pure thought alone.[19]

2.1. *Computation and Mathematical Reasoning*

In speaking of computations, it is important to remember that one may be referring to sequences of mathematical entities, for example, derivations within the Gödel–Herbrand formalism for recursive functions, or instead to physical processes, for example, calculations made with paper and pencil or computer runs. For clarity I will reserve the term 'computation' for the *physical processes* and use the expression 'abstract computation' when referring to computational sequences in mathematical formalisms.

We ought not restrict our attention to just the evaluation of recursive functions using known algorithms. Mathematicians frequently perform computations when they rewrite expressions for complex numbers by applying algebraic rules, evaluate integrals using identities, or manipulate expressions in transfinite number theory by appealing to set-theoretic theorems. Yet most of the functions denoted in these cases transcend the recursive. Such symbol manipulations are computations because they are in principle formalizable as derivations within an effective formalism.

Even the most superficial observation reveals that everyday mathematical activity, the *practice* of mathematics, is replete with (physical) computations. Whether carried out using paper and pencil, computers, or consciously in one's head, these computations often figure as essential pieces of mathematical evidence. An obvious place to see this is in mathematical publications where suppressing computational steps saves pages while challenging readers. Whether an author uses an explicit 'by computation' or an implicit matter of factual appeal to tacit identities, as in 'since $(x + 1/x)^2 = x^2 + 1/x^2 + 2 \ldots$',

[19] My thoughts in this section were prompted by the philosophical discussion of the computer proof of the Four Colour Theorem. See Tymoczko (1979) and responses by Teller (1980), Detlefsen and Luker (1980), Krakowski (1980), and Levin (1981). To a certain extent I travel paths blazed by Tymoczko and Detlefsen and Luker. I have also benefited from conversations with Bijan Parsia, although he denies that the use of computation shows that mathematics is empirical. He sides with Lakatos (1978) in holding that mathematics is quasi-empirical in that it makes non-deductive inferences from mathematical premisses.

readers who want the author's complete evidence must carry out the missing computations for themselves.

Obviously, in so far as computations help prove mathematical theorems, they can also refute mathematical conjectures. Fermat conjectured that all numbers obtained by raising 2 to the nth power of 2 and adding 1 are primes. Computation confirms Fermat's conjecture for $n = 1, 2, 3, 4$, but Euler refuted it by factoring the number obtained when n is 6.

Although a detailed analysis would show the reasoning involved in these illustrations to be quite varied, they share a common element: instead of following the commonplace pattern of using mathematical and physical premisses to draw empirical conclusions, these examples use mathematical and empirical premisses to draw *mathematical* conclusions. This deserves a closer look.

For simplicity let us restrict our attention to physical computations mirroring abstract derivations within a known formalism for recursive function theory. Then the reasoning condensed in a step 'by computation' runs more or less according to the following *Pattern A*.

(1) A reliable human or mechanical computer has carried out a physical process corresponding to deriving '$f(a) = b$' within some formalism S for recursive functions.

Thus

(2) '$f(a) = b$' is derivable within S.

Hence, by the (mathematically demonstrable) soundness of S,

(3) $f(a) = b$.

Of course, we rarely calculate recursive functions within some standard formalism of recursive function theory. In school, for instance, we calculated them using traditional algorithms and decimal notation. Yet specifying a formalism for our computations would be a lengthy but routine task. Hence, it is clear that with some work inferences from those computations could be fitted in the mould of Pattern A.

2.2. *From Empirical Premisses to Formal Conclusions*

In Pattern A the transition from step (1) to step (2) consists in inferring a mathematical claim about a formal object within a mathematical formalism from undeniably empirical claims about the behaviour of a physical device. The other steps use standard mathematical reasoning and do not raise special issues. But transitions of the first kind certainly do; so let us examine them more thoroughly.

The reasoning that concerns us begins with a premiss describing the performance of a reliable human or mechanical computer. One might think that a premiss of this type is not empirical because the concept of computational reliability is a mathematical concept.[20] But I think we can characterize the notion of reliable computer in uncontroversially non-mathematical terms. For a computer to be reliable it must be in good physical (and mental, if appropriate) working condition and properly programmed, and, we might add, produce results which are *generally accepted* as correct. By phrasing the last condition in epistemic terms we avoid using such mathematically defined terms as 'mathematically correct' or 'sound'.

To say that a human computer is in good working condition might be to say that its memory is working well, that it is able to concentrate and follow instructions, that it can read and write legibly, and so on; while to say that a mechanical computer is in good working condition might be to say that its circuits are in good order, that it is in an appropriate physical environment, that its control systems are working, that its input–output devices are ready and working, and so on. Of course, one way to test a whether a human or mechanical computer is in good working order is to give it test computations. This is probably the most efficient way to evaluate human computers, but it need not be the only way. We need only look at the example of electronic computers to see this. Engineers have already developed a wide variety of non-mathematical tests and instruments for determining whether an electronic computer's physical components are in good working order. Admittedly, humans are much more difficult to diagnose than electronic computers, and it may be hard to distinguish between human errors due to mathematical ignorance and those due to temporary or permanent mental or physical disabilities. But even this does not force us to define reliability for human computers in terms the mathematical correctness of their output. For instance, it is well known that we cannot reliably compute with large numbers. Yet we need not bring in the incorrectness of our results to establish this; instead we can point to their

[20] I had such a worry in Resnik (1989), from which the present pages are adapted.

diversity. Taken in groups, reliable computers should give the same answer to the same problem; taken individually, they should give the same answer on repeated trials on the same example.

Of course, even if a computer is in good working order it can reliably execute a particular computation only if it has been correctly programmed to execute the algorithm corresponding to the computation. But again I see no need to characterize a computer's being correctly programmed in terms of the mathematical soundness of its output. Notice that we are not concerned here with whether the program *qua* formal object correctly codes a given algorithm—whether, for instance, a program in BASIC codes Euclid's algorithm. Rather we are concerned with whether the program in question has been properly *installed* in the computer. In the electronic case this can be determined by checking the appropriate contents of the computer's memory. The human case is less straightforward, but presumably asking the computer to describe the steps it would take to execute computations of a given kind would tell us much about whether it had been properly programmed.

Although I think the previous paragraphs make it quite probable that we can specify the reliability of a computer in non-mathematical terms, it might be that some mathematical element remains ineliminable. We might, for example, need statistical measures to describe the uniformity of a group of computer answers to test problems. Yet even if the claim that a computer is reliable must be framed in mathematical terms, the claim itself is an empirical one, since attributing a mathematical concept to an empirical object—calling it, for instance, tripartite or triangular—is to make an empirical claim. Thus premiss (1) remains an empirical one.

This premiss also describes the computer as executing a computation *corresponding to* deriving a formula with a certain formalism. This is to claim that there is a connection between the mathematical situation and the physical one: the computer is supposed to be in physical state P only if the formalism has a corresponding mathematical property M. Where does this connection come from?

There are a number of ways in which one might fill in the connection between state P and M, but I think that they ultimately come to something like this. We can prove mathematically that any system P (physical or otherwise) having certain structural properties can be represented homomorphically by an abstract system M in the sense that (a) there is a one–one composition preserving correspondence between the elements of S and certain elements of M and (b) elements of S have certain structural properties only if their representatives in M do. This allows us to analyse the first premisses of inferences conforming to Pattern A as consisting of an unproblematic mathematical claim:

(1a) Any physical system of type P can be homomorphically represented by a mathematical system of type M;

and two empirical ones:

(1b) The computer in question (in being reliable) is a physical system of type P;
(1c) The computer in question has a certain structural property F.

In short, these premisses amount to claiming that the computer is in a certain physical condition homomorphically represented by a certain configuration in a mathematical formalism. This allows us to infer to the second step stating that the mathematical formalism in question is in the configuration mentioned.

2.3. *What Do Computational Inferences Show Us about the Nature of Mathematical Knowledge?*

Human mathematicians, both theoretical and applied, use physical computations to draw conclusions about mathematical objects. Sometimes they do so deductively in order to prove or refute mathematical claims; sometimes they do so non-deductively to marshal evidence for or against mathematical conjectures. And sometimes the mathematical results they obtain are as theoretical as they come. But does this alone show that important pieces of mathematical knowledge are empirical? Yes, provided that we can set aside some potential apriorist or separatist objections.

The most fundamental of these objections is that in appealing to computations we simply do not reason according to Pattern A. Writing 'by computation' in a proof is like citing a previously proved theorem. We don't appeal to such theorems by reasoning: 'Someone has previously proved that A and I have proved that $A \to B$; so B.' Using a 'by computation' in a proof has the same effect as citing a previously proved theorem; it converts the proof into a proof sketch whose completion requires filling in the missing steps of deduction or computation. Furthermore, when computing we infer the result *directly* at the end of the computation; we don't reason: 'I computed 7×35 and got 245, so $7 \times 35 = 245$.' And the same point applies to proofs assisted by electronic computers: using an electronic computer to do a computation is in principle no different from having a fellow-mathematician do it, and that in turn is in principle no different from doing it oneself.[21]

The objection is correct in that Pattern A does not describe the sort of inferences we run through when giving a proof sketch that includes a step established by a computation (no matter who or what has done it); we simply take the step as a premiss and use it in subsequent inferences. But we should distinguish the reasoning we go through in constructing proof sketches from the justification we would give if someone questioned one of its steps. Since it is practically impossible for any of us to reconstruct all of the mathematics cited in complicated proof sketches, we would have no choice but to appeal to the reliability of the proofs or computations we cited. In either case this would involve empirical premisses, which in the case of computations would conform to Pattern A.

A second objection responds to the point I have just made by differentiating between the methods we actually use in mathematics and those that we would use under ideal circumstances. One might put the objection as follows. Granted, the methods we use are the best for us in our circumstances. (Better to let a computer keep track of my bank balance than to try to do it in my head!) However, *in principle* physical computations are superfluous, since they can be replaced by full-fledged deductions, which take place a step at a time entirely within one's head.[22] At no time would one need to appeal to any physical or mental outcomes. Thus when arriving at the conclusion of the deduction there would be no need to refer to its previous steps or to one's belief in them. Hence mathematical knowledge does differ significantly from scientific knowledge in that any item of mathematical knowledge is in principle deducible from purely mathematical premisses containing no reference to physical or mental events.

The usual way with objections of this sort is to remark that unless one can show that an a priori form of mathematical evidence justi-

[21] Bijan Parsia suggested objections of this sort to me.

[22] Cf. Steiner (1975), ch. 3.

fies the axioms, the epistemic status of the conclusions remains open. This remark is appropriate where evidence for the axioms is wanted, and we have already seen that *ultimately* such evidence is indirect and empirical. None the less, because computations are usually used to establish results *within* axiom systems, one could argue that what we learn from computations is that if the axioms are true then so is the theorem. It would follow that ideal beings would differ from us in being able to justify certain conditional truths via logic alone that we can only justify through using empirical premisses.

A similar point applies to falsifying mathematical results already on the books. Future experience could upset a mathematical proof based upon computations by undercutting our confidence in them. But this would not happen to ideal beings, since their knowledge would be based upon logic alone. Only such experience that could prompt them to revise the laws of logic could upset their proofs.

When philosophers speak of mathematical knowledge as a priori or based upon a special form of evidence, they have in mind at least two intertwined ideas. The first is that it is known independently of experience, the second is that it is immune to experiential revision. I have argued that much of our mathematical knowledge is not special in either sense, for it is in fact based upon experience which for us is largely ineliminable. But I am not denying that some of our mathematical knowledge is grounded in our own deductions from axioms. Thus the difference between us and ideal beings is not that they do, and we do not, have knowledge based upon deduction, but rather that they would have much more of it. This prompts me to ask whether and in what sense knowledge based upon deduction is a priori or different from scientific knowledge.

3. MATHEMATICAL PROOF, LOGICAL DEDUCTION, AND APRIORITY

Many contemporary philosophers of mathematics believe that mathematical proofs are the bastions of a priori mathematical knowledge, because they lead us to logical truths. Specifically, we learn from them that it is logically true that if the proof's axioms are true (or true in a structure) then so is its conclusion. Because these conditional statements are logically true, they are supposed to be insulated from empirically motivated revision. In this section I will argue against this supposition.

3.1. *Mathematical Proof and Logical Deduction*

In Section 2 I appealed to the role of computation in mathematics to argue that most of our proofs are not purely logical deductions from axioms. This would tend to weaken the case that every mathematical proof yields knowledge of a logical truth, because some may be too incomplete to reveal the logical chain linking their conclusions and their premisses. For the rest of this section I want to set aside this issue and others related to my remarks on computation, and make my case independently of them. But before moving on to the main argument in this section I want to remind us that some prominent mathematicians have held that mathematical proofs cannot be reduced to a sequence of logical inferences. Poincaré puts this sentiment eloquently in the following passage:

> [S]hould a naturalist who had never studied the elephant except by means of the microscope think himself sufficiently acquainted with that animal?
>
> Well, there is something analogous to this in mathematics. The logician cuts up, so to speak, each demonstration into a very great number of elementary operations; when we have examined these one after the other and ascertained that each is correct, are we to think we have grasped the real meaning of the demonstration? Shall we have understood it even when, by an effort of memory, we have been able to repeat this proof by reproducing all these elementary operations in just the order in which the inventor had arranged them? Evidently not; we shall not yet possess the entire reality; that I know not what, which makes the unity of the demonstration, will completely elude us.[23]

Thus, according to Poincaré, even after we have checked a proof and have recognized that each step does indeed follow logically from prior steps, we may still have failed to grasp all that the proof conveys. Thus he seems to hold that there is more to learn from a proof than a logical truth.

The intuitionist Brouwer held an even stronger position, namely that logic has no real role in mathematics. While Poincaré insisted that intuition is necessary to grasp the unity of a proof, he also held that logical deduction is necessary to give a theorem its certainty. By

[23] Poincaré (1913), 217.

contrast, Brouwer held that mathematical proofs are mental constructions, and that a theorem merely reports that its author has executed a construction backing it. The being of the construction lies in its being performed. Nothing further in the way of axioms or rules is needed to legitimate it. On this view, the only role logic and axiomatization have to play in the study of mathematics is that of systematically describing—not certifying—the constructions mathematicians have already completed.[24] These descriptions may aid you or me in reconstructing some area of mathematics, but they cannot serve as mathematical evidence. Only our own constructions can do that.

Now I am not endorsing either Poincaré's or the intuitionists' epistemologies. But they do voice a conviction, which many mathematicians hold,[25] that logicians and philosophers impressed by the results of formalization have missed an essential feature of mathematical knowledge. Perhaps one could respond, in the manner of Frege, Carnap, or Hempel, that these mathematicians should distinguish the psychological aspects of discovering theorems from the epistemological aspects of justifying them. But many contemporary epistemologists and philosophers of science could rejoin that such a distinction is no longer viable.

To make the picture even more cloudy, there is no consensus even among the friends of logic as to where logic proper ends and mathematics begins. Some believe that logic ends with first-order logic, and cite the fact that it is the only logic that has a complete proof procedure, is compact, and is subject to the Lowenheim–Skolem theorems as evidence.[26] Others point to the last feature as evidence that the logic of mathematics is not first-order but rather second-order logic.[27] Still others think that logic is given by the ramified theory of types.[28] Furthermore, second- and higher-order logics have no complete proof procedure; their notions of logical consequence are to be understood semantically, and are usually explicated in terms of set theory. On this view of logic, we may even require mathematics to establish some logical consequences![29]

[24] See Heyting (1956), 69–71; Brouwer (1913).

[25] Cf. the disdain the 'ideal mathematician' shows for logic and formal proof in Davis and Hersh (1981).

[26] See Tharp (1975).

[27] See Shapiro (1991).

[28] See Hacking (1969).

[29] The usual practice is to formulate the semantics of higher-order logics within set theory. This would enable us to prove that S is a logical consequence of W by proving a theorem to that effect within set theory. But notice that in order to employ

The only conclusion that I will draw from this inconclusive situation is that it is simply not obvious that in proving a mathematical theorem we (simply) learn through purely logical methods that the theorem is logically necessary given the axioms. For the methods might be both mathematical and logical, and the proof might inform us of some non-logical fact. Still my argument in the remainder of this chapter will not turn upon this. For the sake of argument I will grant that by deducing a mathematical theorem according to accepted rules of logic we can learn that it is true provided the axioms are.

3.2. *Wide Reflective Equilibrium and the 'Epistemology' of Logic*

I deny, however, that mathematical deductions yield a special kind of truth—so-called logically necessary truths. Nor do they guarantee that the only rational grounds we might have for revising the beliefs we base upon these deductions will come from discovering errors in the deductions themselves. Of course, mathematical logic proves that our rules of deduction are formally valid in various technical senses of formal validity. But it does not establish that formal validity suffices for logical necessity or immunity to experiential refutation. If, for example, we take formal validity to mean *comes out true for all interpretations in non-empty universes*, then the claim that a given quantificational schema is formally valid is a claim of set-theoretic model theory. Mathematical logic establishes such model-theoretic truths, but it does not take the further step of connecting them with logical necessities. Indeed, it is not obvious how to make the connection. It is plausible, for example, that set-theoretic models represent logically possible worlds, but this and the claim that the set-theoretic models capture all and only logically possible worlds need arguing.[30]

Consider how one might argue for claims of this sort or for the claim that our rules of inference generate logically necessary consequences. One might start with our logical intuitions (for example, intuitions that certain sentences are logically true) and try to construct an appropriate soundness proof. These intuitions might be either metaphysical ones about various logical necessities or norma-

this theorem to infer that S is a logical consequence of W we must take the axioms of set theory to be true.

[30] Cf. Etchemendy (1990).

tive ones about the logical correctness of various arguments. Either way, the soundness proof will be no better than the intuitions from which it starts. Thus it is unlikely to convince anyone who begins with firm logical intuitions to the contrary. Nor is it likely to convince people, such as myself, who are sceptical about the very notion of logical necessity.

What about turning to Logic, the discipline, as a source of sophisticated methods for justifying claims concerning logical necessities and consequences? The method logicians use when constructing systems for codifying correct reasoning or notions of logical necessity and possibility is the method of wide reflective equilibrium, first identified by Nelson Goodman, but publicized by John Rawls in discussing its use in moral theorizing.[31] One starts with one's own intuitions concerning logical correctness (or logical necessity). These usually take the form of a set of test cases: arguments that one accepts or rejects, statements that one takes to be logically necessary, inconsistent, or equivalent to each. These are—in Rawls's terms—one's *considered judgements*. Given their topic, I find it convenient to call them *considered judgements of logic*. One then tries to build a logical theory[32] whose pronouncements accord with one's initial considered judgements. It is unlikely that initial attempts will produce an exact fit between the theory and the 'data'. Furthermore, committing oneself to a logical theory invariably entails acknowledging unforeseen and prima facie anomalous logical relationships. (The so-called paradoxes of truth-functional implication constitute a case in point.) Sometimes one can respond to such anomalies with a simple modification of one's formal system. Sometimes retranslating a prima facie anomalous argument will reconcile it with the theory. Sometimes, however, one will yield one's logical intuitions to powerful or elegant systematic considerations. In short, 'theory' will lead one to reject the 'data'. Moreover, in deciding what must give, not only should one consider the merits of the logical theory *per se*, such as its simplicity, fruitfulness, or elegance, and the firmness of one's logical intuitions, but one should also consider how the theory and one's intuitions cohere with one's other beliefs and commitments, including philosophical ones. When the theory rejects no

[31] See Goodman (1955); Rawls (1971).

[32] I take a logical theory to consist of a formal system, its semantics, metatheory, and rules for paraphrasing informal sentences into logical notation.

example that one is determined to preserve and countenances none one is determined to reject, then the theory and its terminal set of considered judgements are in, to use Rawls's term, *wide reflective equilibrium*. The equilibrium is wide, because the theory is consonant not only with one's terminal set of considered judgements, the mark of narrow reflective equilibrium, but also with one's broader system of beliefs.

Some discussions of reflective equilibrium suggest that it is a kind of intellectual contentment, a psychological condition characterizing individual logicians or teams of them, and something to be determined more by introspection than by logic. But I mean reflective equilibrium to apply to systems consisting of beliefs, logical theories, and considered judgements of logic. Such a system is in such a state just in case it is coherent by the lights of its own logical theory.

Of course, one's own logical theory, momentary beliefs, and logical intuitions may be in reflective equilibrium only temporarily. New experiences might produce beliefs or logical intuitions conflicting with one's logical theory. Such is our psychological predicament, but this does not make reflective equilibrium a psychological matter.

Since in determining reflective equilibrium one uses the logic contained in one's own evolving logical theory, one might think that a theory may be in reflective equilibrium from its own internal point of view and not so from the point of view of another theory. I hesitate to draw this conclusion, since I wonder whether one could make sense of a rival logical theory while remaining true to one's own. (The problems with interpreting intuitionism illustrate the difficulties one might encounter.) Reflective equilibrium may be a notion that is immanent to a logic rather than transcendent.

Since constructing a logical theory involves balancing various values against each other and making choice after choice, there is no reason to expect it to lead different users to the same outcome. Even if we could make sense of an ideal limit of logical enquiry—when all the 'data' are in—we cannot avoid the possibility of different logicians (or different communities of logicians) correctly applying the method to the same initial data and arriving at different outcomes.

Furthermore, I do not think we can make sense of an ideal point where logicians are bound to agree at least concerning the logical data. Scientists in differing circumstances are likely to begin with different observational data, but it is at least arguable that their opinions concerning the observational data can be made to converge by exposing them to sufficiently similar experiences. Logicians, like scientists, are likely to start theory construction with different initial data. Some will see the derivation of the principle of mathematical induction as a logical deduction, others will argue that it involves set-theoretic reasoning. What some will see as unexceptional uses of excluded middle, others will see as fallacies. But, unlike the scientific case, we cannot contrive for logicians to concur concerning the 'data' unless we do some fancy brainwashing. For it is not just a matter of seeing that they have similar experiences; rather it is a matter of making them come to the same evaluations. Thus *they could even come to agree on all truths not containing expressions such as 'logically true' and 'consistent' while still disagreeing on the extent of logical truth*.

When logical theorizing reaches an equilibrium point the result consists of a logical theory in wide reflective equilibrium with a set of considered judgements of logic. If I am correct about the wide reflective equilibrium methodology, then different logicians can apply the same methodology and arrive at different terminal sets of considered judgements.

Notice that it is common for logicians to develop different formalisms for classifying the same set of arguments and statements. There are complete and sound axiom systems, tree systems, and natural deduction systems for first-order, intuitionist, and modal logic; and while there have been many debates concerning the correctness of each logic, they have never focused upon the type of system used to formulate them. Thus, unlike the scientific case where disagreements commonly arise at the level of theory, disagreements in logic arise at the level of considered judgements. For this reason when I speak of *a logic* I will mean the classifications yielded by a logical theory in reflective equilibrium rather than the formal apparatus of such a theory.

3.3. *Against Logical Realism*

In discussing logical necessity and related notions it is important to distinguish our informal and philosophical notions of logical necessity and possibility from the various types of formal validity, consistency, and implication developed in mathematical logic. My target in this section is realism concerning the informal and philosophical notions, and not realism concerning the latter—which may well be a species of mathematical realism. I will be criticizing *logical realism*—the doctrine that statements attributing logical properties or relationships, such as '"0 = 0" is logically true' or '"0 = 0" does not imply "1 = 0"', are true or false independently of our holding them to be true, our psychology, our linguistic and inferential conventions, or other facts about human beings.

If the previous description of the methodology of logic is correct, then we have no reason for thinking that one logic rather than another has captured the pre-systematic notions of logical necessity and logical correctness. But this raises an epistemological problem for logical realism. To borrow from Benacerraf's criticism of mathematical realism, if logical necessity is a metaphysical property of sentences or propositions, then we have no grounds for thinking that we can always know where it applies.

I do not infer from this that we should turn from realism to logical relativism. We should take our own logic seriously. For suppose that my logical theory, data, and other beliefs are in wide reflective equilibrium—at least according to my own logic. Then no fan of a rival logic can undermine my logical theory by pointing to a defect in it or to anomalous 'data' or to some other problem that both of us have already considered. I have already taken account of these in achieving equilibrium. By my lights I am right and my rivals are wrong—they accept invalid inferences or reject valid ones. (Of course, someone could upset my equilibrium by presenting me with new considerations, but I do not have that in mind now.) Thus there is no reason for me not to take my own logic seriously and no reason for me to tolerate rival logics (other than social or political reasons). This intolerance is not incompatible with my commitment to wide reflective equilibrium. True, but for a whim I would be in the shoes of one of my rivals. But I am not in their shoes. To convert to one of their theories I must either give up intuition *X* or principle *Y*. Why should I do that when my theory is working so well now? From my present point of view, I made the right choices—by luck or whim, perhaps—but that makes them no less right.

On the other hand, this would be no reason for realists (or even myself) to *identify* logical necessity and correctness with the output of our logical methodology. After all, even if we recognize such things as logical necessity and correctness, we should recognize that our procedure may not have produced the true logic. Nor do I think that there is any point in identifying logical necessities with those truths whose denial would violate our linguistic conventions, or with those which belong to our most global theories or are otherwise central to our intellectual life. True, the so-called logical truths, inferences, and equivalences play central roles in our reasoning and theorizing. And persistent logical deviance is a good indicator of some sort of linguistic incompetence. But we should not identify logical facts, if there be any, with facts that might explain why we make the judgements of logic we do. For example, it is hard to believe that in calling a statement logically true, we are describing its role in our conceptual scheme or stating a convention.

Logical realism is, I suspect, the natural inclination of many philosophical logicians just as platonism is the natural inclination of many working mathematicians. However, the case for it is much weaker than that for mathematical realism. Although I will give several reasons why one would be inclined to be a logical realist, in the end each either fails to be compelling or else does not support realism alone.

One of the strongest reasons in favour of logical realism is that matters of logical correctness seem to be independent of our personal wishes and beliefs. What better confirmation of this is there than discovering that one of one's pet arguments is invalid! Surely a philosophy of logic must take into account this bit of phenomenology, and realism about logical values backed by realism about logical properties and relations does it well enough. But the view that logic reflects our linguistic conventions, or the view that logical truths are those that are central to science, and even certain versions of the view that our logic is a function of our psychology, can also ground logic in something independent of our personal opinions, and thereby explain the objectivity (better, intersubjectivity) of logic and the appearance of an independent logical reality. Moreover, they can do so just as plausibly.

A second, related reason is that, by claiming that we are all responding to the same logical values, realism explains the widespread agreement one finds concerning the judgements of logic. But, again, pointing to shared conventions, psychologies, conceptual frameworks, or inferential practices can explain agreement on matters of logical correctness as easily as realism. On these views, you and I agree that Russell's Paradox leads to an absurdity because all humans are inclined to react this way or because we share the same language or inferential practices, and so forth. Furthermore, pointing to these can better explain the significant disagreements that exist.

A third reason that might come to mind is that much of contemporary philosophy, especially metaphysics and philosophy of language, contains theories built upon a realist view of logical possibility, necessity, implication, or equivalence. This reason is thus analogous to indispensability arguments for mathematical realism. Now the indispensability arguments for mathematics are predicated upon our use of mathematics in science, and we dismiss mathematized science at our peril. But it can hardly be said that we dismiss any contemporary metaphysics or philosophy of language at our peril, much less possible-worlds semantics; so I do not find much force in this indispensability argument for logical realism.

In conversation Keith Simmons has suggested another reason, namely that only realism about logical facts can explain our intuition, for example, that if we used a different logic or even if no rational beings existed at all, it would still be the case that no number is both even and not even. Like the last reason this by itself is no argument for realism about logical values. I see three ways to interpret it, and several ways for an anti-realist to respond. First, taking the objection at face value, anti-realists can prove in the usual way *using the logic we now have* that under no condition would a number be both even and not even, and that this is why none would be so if we had a different logic. The point is that changing our logic would not alter the truth of these statements, but at most our attributions of truth to them. To be sure, anti-realists cannot explain why it would still be contradictory for a number to be both even and not even if we had a different logic, but this *explanandum* is itself a thesis of logical realism. However, I think that Simmons had in mind something like the idea that even if our conventions for 'no', 'and', and 'not' were different it would still be true that no number is both even and not even. Now if saying that 'it is true that *p*' is just another way of saying that the sentence supplanting '*p*' is true, then the intuition in question is wrong. Without specifying the case further, there is just no saying what the truth-value of 'no number is both even and not even' would be if we used the words 'no', 'and', and 'not' differently. On the other hand, if 'it is true that *p*' is an operator applying to propositions or Fregean thoughts, then anti-realists can use our current logic to argue as before that no matter how we used our current logical vocabulary it would still be true that no number is both even and not even.

I also suspect that realists about logical necessity and other properties and relations often confuse normative intuitions concerning logical correctness with descriptive ones concerning metaphysics, and tacitly base their descriptive metaphysics on normative judgements of logic. If so, we can add to the list of problems for logical realism the problem of showing why normative intuitions are relevant to metaphysical facts.

Let me reiterate that my anti-realism about logical values or logical properties and relations does not extend to anti-realism about the claims made in formal logic. For example, I have no objection to being a realist about whether sentences in a sufficiently regimented language are truth-functional tautologies or whether one is a first-order consequence of others. Holding that a sentence is true under all interpretations in every non-empty universe commits one to its being true, but not to its belonging to a special category of logical truths in either the evaluative or the metaphysical sense.

Before I pass on to propose my own anti-realist account of logic, I want to emphasize that the objections to logical realism—whose similarity to those against mathematical realism is striking—spell trouble for philosophies of mathematics which trade mathematical facts for logical ones.[33] The contemporary version of this movement originates with Hilary Putnam, who proposed that we view talk about mathematical objects and talk about the logical implications of the axioms of a branch of mathematics as 'equivalent descriptions' of the same mathematical facts, and use one viewpoint 'to clarify the other'. The problems in the philosophy of mathematics

[33] The objections are so similar to those to mathematical realism that Pieranna Garavaso wonders (in correspondence) why I don't accept the force of the latter. First of all, I do accept the anti-realist arguments based upon pointing out that we can give non-realist accounts of many of the facts about mathematics, such as its apparent objectivity, which realism claims to explain. These show that realism cannot claim to be the only explanation of these facts. But I don't accept the objection that we lack an epistemology for mathematical realism, for the obvious reason that I think I provide one in this book. And while I think that we need not appeal to statements of the form '. . . is logically true' or '. . . is logically consistent' in doing science, I do not think that we can dispense with mathematical objects and truths in doing science.

arise simply because we become captivated by just one picture of mathematics. Here is how he put it:

> In short, if one fastens on the first picture (the 'object' picture), then mathematics is wholly extensional, but presupposes a vast totality of eternal objects; while if one fastens on the second picture (the 'modal' picture), then mathematics has no special objects of its own, but simply tells us what follows from what. If 'Platonism' has appeared to be the issue in the philosophy of mathematics of recent years, I suggest that it is because we have been too much in the grip of the first picture.[34]

Putnam seems to regard either picture of mathematics as appropriate and useful, but his students Hartry Field and Geoffrey Hellman have drawn upon his insights to further their own versions of anti-realism. Field rejected mathematical objects in favour of the view that mathematical knowledge is really knowledge of logical necessities and possibilities. (His modal picture of mathematics differs from Putnam's in ways that need not concern us here.) Hellman, on the other hand, carefully filled in the details of Putnam's sketch of a modal-structuralist picture of mathematics, but then departed from Putnam by rejecting the mathematical objects picture of mathematics in favour of the modal picture.

Both Field and Hellman are realists about logical necessity and possibility, and they take these notions as primitives which may not be defined using set-theoretic models or possible-worlds semantics. Now if logical realism is untenable, if there are no logical facts, then, contrary to Putnam, we don't have two equally satisfactory pictures of mathematics, and contrary to Field and Hellman, their modal pictures cannot substitute for the mathematical objects picture. Furthermore, even if mathematical claims are equivalent to claims in formal modal logic, further argument is needed to show that this modal logic has factual content (other than the mathematical content furnished through interpreting it via possible-worlds semantics).[35]

3.4. *In Favour of Logical Non-Cognitivism*

The force of my criticisms of logical realism don't depend upon providing a substitute for it, but in this section I want to take steps

[34] Putnam (1967), 330.

[35] My criticism has no force, of course, against those who would view neither mathematics nor modal logic as making factual claims.

towards doing so. The view I am proposing is a restrained *logical non-cognitivism*: sentences of ordinary language that seem *categorically* to attribute logical necessity or other logical properties and relations actually perform other functions, and are neither true nor false.

The view concerns sentences of ordinary language containing such terms as 'implies', 'consistent', and even 'logical' (as in 'that was the logical thing to do'), but not sentences containing such terms as 'logical truth', 'logically inconsistent', 'logically impossible', which are philosophers' terms of art.[36] I will not offer an account of what philosophers might mean when they use the latter terms. I doubt that a single account will work, since philosophers differ from each other concerning the nature of logic, and, presumably, give different senses to terms like 'logically true'. I did argue in the last section, to be sure, that the logical realists' use of this term commits them to an unacceptable view.

While philosophers introduce the terms 'logically inconsistent' and 'logically implies' in the course of constructing philosophical theories, they clearly intend the terms as refinements or explications of our ordinary terms 'consistent' and 'implies'. Thus my non-cognitivist account of the latter terms has implications for philosophical attempts to explicate them.

The words 'hence', 'therefore', 'whence', 'thus', 'consequently' are used to indicate that an inference is being made. Given their grammar and function, there is little temptation to see them as expressing some property or relation. We need to look instead at terms that are not functioning as inference words.

Consider then sentences, such as the following, which we use to make assessments of a logical character:

> Your remarks are consistent with what you said yesterday;
> That does not follow from the stated conditions;
> By implication the ruling excludes this use too;
> This is an equivalent version of the doctrine.

On the version of non-cognitivism I favour, judgements of this sort serve at least two functions. First, they play a signalling role in our inferential practice. In calling someone's remark 'contradictory', for

[36] I am indebted to Eric Heintzberger for bringing this point to my attention and prompting me to revise this section.

example, we are not describing it as, say, false under all substitutions for its non-logical terms or as false in all possible worlds. Rather we manifest that we expect it to be treated in a certain way. Among other things, we show that we are confident that our audience can see this for themselves, that the person uttering the remark should retract or qualify it, and, perhaps, that we will regard those who persist in failing to see our point as intellectually incompetent. (Compare 'it's obvious'.) Categorical judgements of consistency, implication and equivalence play a similar role in our inferential practice by policing transitions between statements and commitments to groups of statements.

A second function of these judgements is to express our commitment to certain factual claims. For example, in saying that *A* is trivially true, I commit myself to its truth; while in saying that it implies *B*, I commit myself to its being true if *B* is; and in saying that it is inconsistent with *B*, I commit myself to the falsity of their conjunction. By saying these things I open myself to criticism if the claims to which I commit myself turn out to be false.

We can even see these commitments as matters of implication, and judge, for instance, the inconsistency of *A* to imply its negation, so long as we see these judgements on a par with other judgements of implication. Thus saying that the inconsistency of *A* implies its negation is to signal both that the transition from the first to the second is permissible and that whoever claims that *A* is inconsistent is committed to the truth of its negation. But when we say that the inconsistency of *A* implies its negation we do not describe the two as standing in a logical relationship of implication in any stronger sense.

Although I deny truth-values to sentences expressing categorical judgements of logic, I see no reason to refrain from treating them as if they have truth-values in developing their 'logic'. This is what anti-realists about possible worlds do when they systematize modal inferences using the notion of validity defined in a modal logic via an index set *W* and a possible-worlds semantics. Here remaining true to one's modal anti-realism is a matter of not taking talk of possible worlds as attributing genuine truth-conditions to modal sentences. Similarly, so long as we refrain from assigning factual truth-conditions to judgements of logic, developing a 'logic'—even a two-valued logic—using a set of indices which we might, for convenience, call 'truth-values' would not betray non-cognitivism.

To my mind, the most serious objection to non-cognitivism, whether the view concerns terms of logic, modal language, ethical discourse, or discussions of rationality, is that these terms seem to function descriptively in many sentences that appear to have truth-values. Here are some examples using 'imply' and 'contradict':

(1) Since Quine accepts theories that imply that there are mathematical objects, he is committed to them.
(2) Any theory implying a falsehood is false itself.
(3) Frege's earlier views on meaning do not contradict his later ones.

The problem is that these statements have truth-values, and it is hard to see how they could unless 'imply' and 'contradict' can be true or false of certain pairs of objects. Now I think we can acknowledge that terms of this sort are true or false of something in non-categorical contexts without giving up the general thrust of non-cognitivism. One way to do this is to interpret statements like these as tacitly referring to the norms that govern (or ought to govern) our inferential practices. Thus we might render (1), (2), and (3) as:

(1′) Since Quine accepts theories that (according to our inferential practice) commit him to its being true that there are mathematical objects, he is committed to them.
(2′) Any theory from which we may infer a falsehood is false itself.
(3′) Frege's earlier views on meaning do not prohibit one from asserting his later ones.

One might think that this ploy commits one to logical relativism, but distinguishing categorical from embedded judgements avoids this. A logical relativist holds that all judgements of consistency and implication are relative to some background inferential practice, and that statements of the apparent forms '*A* implies *B*' and '*A* contradicts *B*' are actually of the forms '*A* implies *B* in *S*' and '*A* contradicts *B* in *S*', where '*S*' stands for an inferential practice or the 'logic' of an appropriate community. I hold nothing like this concerning categorical judgements, and my view concerning embedded judgements is not as strong. For I do not claim that they have different forms from their apparent forms.

My treatment of (1)–(3) is clearly no more than a start on the problem of accounting for examples of this type in terms acceptable to non-cognitivists. I would prefer to have the sort of systematic treatment Simon Blackburn and Allan Gibbard have developed for embedded judgements of necessity (Blackburn) and rationality (Gibbard), and, perhaps, one of these can be tailored to judgements of logic.[37] At this time I will simply say that neither of them fully satisfies me, and I still regard the problem of embedded judgements as a serious, though not fatal, difficulty for my view.

Someone might worry that in denying truth-values to categorical judgements of logic I undermine the very practice that I claim they help regulate. Perhaps considering the following dialogue can set this worry aside.

> I: *A*, and if *A* then *B*; so *B*.
> You: Really? Why?
> I: Well, if *A* is true and true only if *B* is, then *B* is true.
> You: Yes, but why if that, and if *A*, and if *A* only if *B*, does *B* follow?
> I: (to stop the Carrollian regress) What do you mean by 'Why does *B* follow?'? If you mean, 'Why is *B* true?', then I have already given you the reason. If you mean 'Why should I accept *B*?', then I answer: because it is true.
> You: But why should I accept the truth?
> I: Because you just should!

In this kind of a predicament we must eventually stop stating reasons or rules, and simply try to persuade or compel our audience to follow our practice. Adding the response, 'Because *B* is a logical consequence of *A* and "if *A* then *B*",' will not extricate us; it will only postpone resorting to pleas, exhortations, threats, and cruder forms of persuasion.[38]

One might also wonder whether as a non-cognitivist I can consistently endorse wide reflective equilibrium or even the theoretical discipline of logic. I do not see any inherent conflict here. Our informal judgements on matters of logic help to promote, inculcate, and enforce our inferential practice. But considerations of system play a

[37] See Blackburn (1984); Gibbard (1990).

[38] One way to persuade a reluctant audience is to teach them formal logic and carry them through the process of wide reflective equilibrium. Then we will be able to state as a further consideration that *A* and 'if *A* then *B*' truth-functionally imply *B*. But in so doing we will simply describe these statements in technical terms and will not affirm a judgement of logic concerning them.

large role in this very practice, since the practice governs what we may or may not hold while continuing to maintain our prior commitments.[39] System is not only a concern of this practice, it is essential to its success. Thus it is entirely appropriate for us to use this practice to systematize and criticize it, which is just what we do in using wide reflective equilibrium.

Finally, one might wonder, as Mark Balaguer has in correspondence, whether my position on logic undercuts my mathematical realism. I think that it should be pretty clear that it does not. First, I deny no truths that mathematics affirms. I do not deny, for example, that either there are or there are not infinitely many twin primes; I simply refrain from affirming this as a logical truth in any realist sense of logical truth. Nor have I called for revisions in the deductive methodology of mathematics; I retain all the connections mathematicians recognize between their axioms and theorems. What I deny is that these connections hold by virtue of an objective logical relation between them, that is, a relation that holds independently of our beliefs, inferential practices, and the like.

3.5. *The Apriority of Logic*

In proving a mathematical theorem we do not come to know a logical fact (for instance, that the axioms logically necessitate the theorem) which holds independently of our beliefs, proofs, and practices. However, we do come to know the conditional truth that axioms hold only if the theorem does and that the axioms and theorem are connected via deduction. This permits us to insulate the truth from empirical refutation.

It is essential that we have such a practice and truths that we treat in this way. For theory development and testing must take place against a background of principles and rules for generating consequences and commitments. What we call our logic is what we take as fixed in testing and developing our theories.

But in using logic in this way, we place it outside the circuit by which we test our beliefs, and thereby allow no experience to confirm or refute logical truths. So in this sense—of being outside the circuit through which we test our other beliefs against experience—logic is a priori. Consequently, so is our knowledge that a theorem

[39] Notice that one need not use such terms as 'logically follows' or 'consistency' in describing the systematic focus of our inferential practice.

follows from the axioms, if that knowledge has its source in a logical deduction.

Despite this, even logic is not a priori in another sense, that is, of being immune to revision. We might find that certain developments in science and mathematics generate tensions in our system of beliefs so disquieting and so difficult to resolve that we begin to look at the framework that links the system together and to experience. Thus we might find it rational to revise our logic or its limits—to alter our inferential norms, that is—and in so doing change what we count as logical truths and implications. We can certainly find such proposals in the history of science and mathematics. Both the intuitionists and the founders of quantum logic have argued that different logics are appropriate to different intellectual domains. Nor need we expect that the move will always be to curtail or subdivide logic. From the point of view of nineteenth-century logic the movement originating in Frege, and continuing with contemporary fans of higher-order logics, has promoted an expansion of the limits of logic.

I am not an enthusiast of all of this expansion, being inclined instead to treat so-called higher-order reasoning as part of mathematics. As I see it, the issue here is a practical one. In crediting something's truth to logic we do confer methodological honours and protections upon it. We thereby give it a certain role in ongoing enquiry. In view of this, clarity and predictability would favour limiting logic to what can be captured within an effective proof methods. This favours first- over higher-order logics. In calling something logical, we tend to erect barriers to future inquiry concerning it, and limit the options we are prepared to consider. In limiting our options we should be conservative. Other things being equal, our logic should be minimal.

4. SUMMARY

In the past three chapters I have been arguing that there is no sharp ontological or epistemological boundary between mathematics and the rest of science. The differences between numbers and trees are clear enough, but those between abstract mathematical objects and virtual processes or electron fields are sufficiently elusive to undermine the supposed ontic division between mathematical objects and physical ones. Furthermore, empirical evidence bears upon theories that are combinations of mathematical and 'empirical' hypotheses. Considerations of good sense or pragmatic rationality, rather than reasons of logic, justify our practice of applying a local conception of evidence—one that takes our observations to bear upon the hypotheses of the local science instead of the mathematical ones belonging to a more global theory. Furthermore, the rule governing this practice is the general rule against tinkering with the global theories one's local hypotheses presuppose. The same rule that leads physicists not to question mathematical results also leads biologists not to question established physical or chemical principles. The relative apriority of mathematics is thus due to its role as the most global theory science uses rather than to some purely logical considerations that shield it from experiential refutation. The same good sense counsels us to use Euclidean rescues to save our mathematical hypotheses from empirical refutation, that is, to save them by holding that a putative physical application failed to exhibit a structure appropriate to the mathematics in question. Thus by using a pragmatic view of the methodology of science to temper the holistic account of the relationship between theories and empirical evidence, we can not only account for but even endorse the relative apriority that in practice mathematics enjoys.

It is interesting, then, that even the conception of evidence that is local to mathematics does not entirely exclude empirical evidence. Our experience working within a mathematical theory and in applying it can support our belief that it is consistent. Empirical evidence may also be used to establish the reliability of our computers—something we presuppose when we appeal to their computations. Finally logic is a priori only in a weak sense. It does not report special logical facts and relationships that are known or discerned in a special way. Rather, it delineates a set of truths and relationships which we have for now marked as immune to refutation on the grounds that they are a convenient way to define the parameters through which confirmation and refutation takes place.

There are still some questions about the local conception of mathematical evidence that I have not addressed in this chapter. Why don't mathematicians try to detect the objects they posit? And related to this: why don't mathematical objects themselves play any role in the practice of mathematics although symbols for them do? To answer these questions we need to have a fuller account of the nature of mathematical objects. After we have seen such an account in Part Three we can return to these epistemological questions.

9

Positing Mathematical Objects

1. INTRODUCTION

In discussing Benacerraf's objection to mathematical realism (Chapter 3), I concluded that a realist epistemology need not be founded upon causal or information generating transactions between humans and mathematical objects. I granted, however, that, in so far as realists maintain that mathematical objects are causally inert and outside space-time, they should explain how we can attain mathematical knowledge using just our ordinary faculties. I will now attempt to meet this challenge through a postulational account of the genesis of our mathematical knowledge.

The basic idea is that humans brought mathematical objects into their ken by positing them. Now to *posit* a new kind of object one need only introduce a new predicate *P* (or, as happens frequently, begin to use an old one with a new sense) and claim that *P* exists. Thus, it is plain that realists who claim that mathematical objects are posits invite a variety of worries and objections. Postulational approaches seem better suited to conventionalists, who may claim that we make truths, than to realists, who must hold that we can only recognize independently obtaining truths. Below I will explain why positing is not incompatible with realism, I will distinguish positing in mathematics from the creative efforts of fiction writers, and, finally, I will explain how in positing mathematical objects we manage to refer to them.

I will be assuming that in providing an epistemology for mathematics, realists are entitled to assume that we already have an abundant fund of knowledge of mathematical objects. For the mathematical epistemologist is no more obliged to establish the existence of the mathematical realm than the epistemologist of common-sense knowledge is obliged to establish the existence of an external world. The challenge for both is to explain how we acquire the knowledge they assume we have.[1] Of course, the mathematical epistemologist has the prima facie more difficult problem of explaining both how we acquire knowledge of a realm with which we cannot interact and how anyone managed to acquire any initial mathematical knowledge.

The problem, then, is to explain how we have obtained the knowledge of mathematical objects we now have. The problem is relatively tractable when it comes to accounting for how we generate new knowledge about those mathematical objects already known to us. We have at least a rough understanding of what goes on: starting with information about these objects that we already have, we manipulate formalisms or diagrams representing this information, or appeal to analogies with other mathematical theories in order to generate new mathematical beliefs or hypotheses. If our methods fall short of justifying our beliefs according to accepted mathematical standards, then we can go on to try deriving them from previously accepted results. This method for generating new mathematical beliefs from old ones is similar to the methods scientific theoreticians use to generate new hypotheses, and neither requires the researchers using them to interact with the objects the new beliefs concern.

We also have some information about how certain historical and contemporary mathematicians came to believe in new types of mathematical objects. Again, it is significant that they have been guided by analogies with previous established mathematics instead of a priori insights. Furthermore, while we may not fully understand the methods these mathematicians used, we may be confident that we can learn more about them through psychological and sociological investigations of contemporary mathematical practice and studies of historical mathematics.[2]

Unfortunately, studying the history of post-Greek mathematical discoveries is unlikely to tell us much about how people without a mathematical heritage originally attained mathematical knowledge. For the mathematicians in question already knew quite a bit of mathematics and its history, and thus they were prepared to believe

[1] In making this assumption I am merely falling in line with the tradition in contemporary epistemology initiated by Quine (1969b). Of course, I have already argued for the recognition of mathematical objects in Chapter 4.

[2] See Maddy (1988).

in new mathematical objects and to compare themselves to previous mathematical innovators.[3] To make matters worse for us, the historical record of very early mathematics is scant, and it is totally mute on the question of how people initially attained knowledge of the mathematical realm.

2. A QUASI-HISTORICAL ACCOUNT

Despite our lack of historical knowledge, I think we can tell a plausible story about the origins of the belief in mathematical objects by interspersing speculations in the historical record.[4] This is what I shall attempt in this section.

To begin, we know that all aboriginal peoples have terms for describing the size of very small collections. Depending upon the particular group, they are likely to have words we would translate as 'one', 'two', 'five', or 'many', or 'very many'. Furthermore, although they have no systems of numerals, they generally have developed means for counting and recording the size of larger collections, such as notching tallies on sticks, knotting strings, bagging pebbles, and the like. Consequently, it would be reasonable to suppose that they have a term roughly equivalent to our 'equi-numerous', through which they at least implicitly recognize that equi-numerosity is symmetric, transitive, and reflexive. It would also be reasonable to suppose that they have terms for comparing sizes and know that one can determine whether one collection is smaller than another by comparing tallies. By appealing to the analogy between contemporary aborigines and the prehistorical ancestors of early mathematicians, we can fairly attribute similar knowledge and abilities to the latter.

It is also plausible that the prehistorics reached a similar stage of sophistication in dealing with shapes—namely, that they had terms for various shapes, and the concept of two things having the same

[3] It is often said that mathematicians use a realist working philosophy which they quickly abjure when pressed with philosophical questions. But I wouldn't take this as evidence that mathematicians do not believe in mathematical objects: first, because for every closet formalist there is a dyed-in-the-wool platonist; secondly, because it is idle to wonder whether mathematicians who willingly affirm that, say, there are infinite numbers actually believe that there are.

[4] For more on the early development of mathematics see Wilder (1968).

shape. Perhaps they also had proto-geometrical knowledge, for example, that a rectangular-shaped thing can be divided into square and triangular pieces. (This is something an observant person could pick up in a day of watching carpenters frame a wooden house.)

The next step for people who have come this far would be to develop indefinitely protractible systems of numerals for counting and diagrams for representing shapes and methods for constructing the diagrams of some shapes from those of other shapes. It is unlikely that these came before the development of written languages, and they probably evolved with writing. As far as I know, nobody has observed contemporary aboriginal people making the transition to this stage (except, of course, through learning it from other cultures), and we have no record of how the ancients made the transition to this stage. Indeed, the earliest documents from ancient Egypt, Babylonia, India, and China already show that their peoples possessed full notations for at least the positive natural numbers and were able to solve simple arithmetical and word problems.

Notice that one can get this far without having to recognize numbers in their own right. (In Quinean terms, one need not quantify over numbers.) One might even affirm that given any *number of things* there might be more, or that there is no limit on the size of the universe, without thereby committing oneself to the existence of infinitely many numbers.

It is not certain that the precursors of the Greeks—the Babylonians and Egyptians—recognized numbers in their own right. On the one hand, they did develop potentially infinite systems of numerals, and they had algorithms for solving numerical problems, which suggest that they at least implicitly recognized the number series. On the other hand, their algorithms can be interpreted as rules for symbol manipulation; and it is not clear that they formulated laws with sufficient generality and abstraction to commit themselves to numbers.

With the Greeks the recognition of mathematical objects as abstract entities is beyond doubt. It is evident both in the language of their theories, and in their philosophical commentaries. Unfortunately, we do not know what led to the transition from the ontically neutral Egyptian–Babylonian stage to the ontically commissive stage of the Greeks. So let me speculate a bit.

Prehistorical peoples, I have assumed, knew that three things are fewer than five things, and they probably learned this through experience in counting and comparing. Yet they could hardly have known that one billion and three things are fewer than one billion and five things. For without a system of numerals one cannot even formulate this claim. Furthermore, it is humanly impossible even for someone with a system of numerals to learn this fact by counting and comparing collections of the appropriate size (or by using logic to show that the existence of one billion and five things implies that of one billion and three things). But it is reasonable to suppose that the Egyptians and Babylonian mathematicians certainly knew things like this. For they could have discovered it easily by reflecting on the norms defining their system of numerals and its use in counting. If they had, they would have seen that one counts their analogue to a '—— 3' before counting '—— 5' and, consequently, that the former counts a smaller subcollection of what the latter counts. Notice that this example also indicates how deductive reasoning and proof could emerge as a method of mathematical discovery and justification. In fact, ancient Babylonians and Egyptians had already developed the practice of proving solutions to word problems by verifying that they satisfied the conditions defining the problems in question.

So far I have argued that it is plausible to interpret pre-Greek mathematicians as knowing how to devise systems of notation, defined by rules of construction and manipulation, which they used to solve practical problems of measuring, counting, and building. Here they would be working in the 'object language'. But, as we have just seen, they must have had 'metalinguistic knowledge', in so far as they were aware of various properties of their notational systems. Of course, none of this forced them to recognize mathematical objects or abstract entities. (Or, if you will, it does not require us to interpret them as committed to such objects.) When speaking of numerals they could take themselves to be speaking of tokens of numerals or the possibility of tokening them. Probably we could construe their proto-geometries as similarly concerned with token diagrams rather than abstract shapes. Thus we needn't wonder how they could have come this far without interacting with abstract mathematical objects.

Yet, clearly, at this point the ancients were at the brink of recognizing abstract entities. Perhaps they just unconsciously slipped into talk committing them to abstract mathematical objects. Our language lets us do this quite easily. By 'a square' we might mean a concrete square-shaped thing or an abstract geometrical object.

Similarly, we might use 'word' to designate a token or a type. What is more, our word 'number' can be used to signify a concrete collection, as in 'A number of people asked for you', a mathematical object, or even a numeral type or token.

Thus the switch from talk of tokened numerals to abstract numbers or from shaped things to abstract shapes may well have come about unnoticed and unconsciously, and became well-entrenched before anyone realized it. I will refer to this way of introducing mathematical objects as *implicitly positing* them. I think it is useful, however, to pretend that the ancients posited mathematical objects explicitly, because it will prompt us to consider the sorts of objections they might have encountered and the responses they might have given. This in turn will furnish us with a better understanding of the factors that might have unconsciously motivated those who implicitly posited mathematical objects. So with this in mind I will continue my story.

Without having to recognize mathematical objects the ancients would have observed that some objects are straighter or rounder or squarer than others. This could have led them to understand what it is for something to be perfectly straight, or perfectly round, or perfectly square. And they would surely appreciate the benefits to be gained from formulating geometric principles and rules of construction in terms of such ideals. For example, butting two square corners produces a straight line but butting two corners that pass for square need not produce a line passing for straight. One the other hand, while we can see why the ancients might have reason to speak of geometric ideals, we can also see why they would have had reason to doubt that such objects can be found in the material world. For, amongst other things, they would have observed that often what appears to be perfectly straight fails to be upon closer examination.[5]

Other considerations might have prompted them to distinguish between what is and what might be, that is, between the actual and the (merely) possible. For example, they might have drawn 'possible' carts. But thoughts about unending progressions of numerals or indefinitely divisible lines or areas would have reinforced the distinction. At this point some ancients might have thought to deal with these possibilities through introducing abstract mathematical objects. However, their sceptical colleagues would have shown them

[5] Discussions with Marcus Giaquinto prompted the last few sentences.

how to account for intuitions about endlessly counting or dividing lines in terms of the possibility of performing more and more actions. Of course, such actions would raise the possibility of ever more matter, ever increasing lifetimes and attention spans, ever diminishing marks, and the like.

However, this way of doing without mathematical objects would have come to an end when it came to geometric ideals. Perhaps stretching a cord tighter and tighter eventually forces it to be perfectly straight, but cutting a line into smaller and smaller segments will not yield an extensionless point, nor will drawing finer and finer lines produce one without breadth. Thus the desire to posit geometric limits would have forced the ancients to give up speaking in terms of possible *concreta*, and to recognize points, lines, and circles as *sui generis* entities, existing in their own right.

Another reason for positing geometric ideals would be the gains to be had in overall theoretical simplicity, clarity, and economy. We have already seen that accounting for the possibility of repeatedly dividing or prolonging a concrete line would require positing the further possibility of unlimited time, matter, and so on. We can add to this that accounting for geometric ideals in this way would also require one to explain how one can complete an infinite set of concrete tasks—a feat which we still have difficulty making intelligible today. Thus it would have made more sense for the ancients to reject the idea that geometrical ideals are some sort of actual or possible *concreta* and to posit them as non-material and timeless things to which our concrete objects at most approximate.

Once one thought to posit geometric ideals it would be a short step to positing numbers (or abstract numerals) so as to attain the advantages of an indefinitely long number series. For taking this step would be nothing compared to positing infinitely small and infinitely large geometric objects.

Despite the recognizable advantages of positing mathematical objects over trying to deal with possibilities involving concrete ones, sceptical colleagues of our ancient mathematicians might have asked them how anyone could learn anything about these new entities. In responding, they would not have been obliged to supply a full epistemology; it would have sufficed for them to show that we can learn enough about the new entities for overall mathematical progress to be made. They could have pointed out that reasoning, already recognized as a successful non-observational method of discovery, can be applied to the new entities. Furthermore, they could (and would) also have postulated connections between the new entities and older symbol systems and diagrams. For example, by assuming that drawn triangles have approximately the same structure as abstract ones, they could have used the former to suggest hypotheses about the latter, which they could have attempted to verify deductively. By assuming that initial segments of the unary numeral sequence are isomorphic to initial segments of the natural number sequence, they could have manipulated the former to learn about the latter. What is more, positing connections of this sort would have allowed our ancient mathematicians to recover in their new abstract geometry and arithmetic many of the results they had previously established in their concrete 'geometry' and 'arithmetic'. All this would have amounted to a strong reply to their sceptical colleagues.

3. MATHEMATICAL POSITING NATURALIZED?

The speculations in the last section cite no supernatural processes in accounting for the genesis of mathematical knowledge. But they do presuppose that before the ancients ever posited mathematical objects they had already developed the ability to communicate in written languages, to use pictures, diagrams, and words to represent things that are absent or merely imagined, to speculate, and, finally, to hypothesize and theorize about new kinds of entities. Of course, I tacitly assumed that none of these abilities involve supernatural processes. Now some philosophers have held views that seem to imply the contrary. For instance, Frege held that using a language, reasoning, and thinking required 'grasping thoughts'. Fregean 'thoughts' are abstract, non-mental, non-physical objects associated with sentences as their meanings, and 'grasping' them takes place through non-perceptual contact with the abstract realm. So if Frege is correct, then my account fails to accomplish its purpose of explaining how we could come to know abstract entities by already presupposing that we interact with them.

I happen to think that views like Frege's are wrong. For the idea that we learn anything by supernaturally grasping abstract entities strikes me as a scientific dead end. I suspect that this is why it appeals to so few cognitive scientists or linguists. In any case, I do not have to argue this point here, since the philosophical debate in which I am engaged already presupposes it. Anti-realists maintain that the *abstractness* of mathematical objects generates epistemic problems *in addition* to any that may arise in empirical science. They thereby presume that attaining scientific and common-sense knowledge does not depend upon our ability to know abstract entities.

On the other hand, if philosophers of the Fregean bent are correct, then, of course, there is no special problem with knowing mathematical objects; since we would have already grasped abstract objects in the course of coming to our most ordinary beliefs. Yet this would not dismiss the problem of accounting for the genesis of mathematical knowledge. We also had the ability to interact with oxygen prior to our discovering it. Yet to point this out does not even begin to explain how we came to recognize oxygen or its role in our body chemistry. Thus it would be reasonable to ask Fregean philosophers how anyone could become aware that there are abstract mathematical objects, since we appear to be just as unconscious of using abstract entities for communication and related purposes as we are of consuming oxygen. These philosophers could easily adapt my account to fit their views on language.[6]

The intensional notions that play such a crucial role in my historical account prompt worries from a different quarter. I have speculated about what other people believe, and have made claims about what they might have stated and how they might have reasoned. What is more, I have drawn upon anthropological interpretations of contemporary aborigines and historical accounts grounded in translations of ancient texts. Quine's reservations about intensional notions, such as belief, and his arguments against the determinacy of translation, are as familiar to contemporary philosophy as his naturalism.

While I share Quine's reservations, my position in the dialectic with anti-realists does not require me to deal with Quine's worries head-on. I know of no anti-realists in the philosophy of mathematics who accept Quine's views on intensionality and translation.

[6] For a recent example of work of the Fregean bent see Linski and Zalta (1995). Linsky and Zalta argue that we must posit a rich ontology of abstract objects, more than sufficient for interpreting mathematics, in order to understand the practice and language of science. Note, however, they do not speak of 'grasping' abstract entities or enlist supernatural faculties.

Indeed, many of them embrace modality, which Quine has long regarded as an abomination.

On the other hand, this chapter can be seen as a part of the Quinean project of naturalizing epistemology, and the method I have used is similar to the speculative one he initiated in *Word and Object*, elaborated in *Roots of Reference*, and reaffirmed in *Pursuit of Truth*. One would hope then that ultimately its use of intensional notions could be put on a non-intensional basis. This would not only underwrite my speculations, but also historical accounts more generally and some of Quine's remarks in particular.

Now Quine has suggested that we can identify a person's specific beliefs with states of their body. For him, the tricky problems arise when we try to make sense of attributing specific beliefs to others. On reading some ancient text a historian might report in English, 'The Greeks believed that there are infinitely many numbers.' Reading this I might see two ways of interpreting the claim. This might lead me to wonder whether the Greeks simply believed that for every number there is a greater or whether they believed the more sophisticated claim that the number series cannot be matched with one of its initial segments. The matter might turn upon issues of translation—and more.[7] Because of this Quine denies that there is a fact of the matter as to what the Greeks believed. On the other hand, he does not deny that there is a fact of the matter as to what they believed provided that we relativize our belief attributions to a translation manual and to our purpose in attributing the beliefs to them and to the situation in which we do so. Of course, like the rest of us, Quine usually leaves these to be tacitly understood when attributing beliefs. Presumably, both his own attributions and my historical account can be reconstructed in the relativized fashion he approves. (And, presumably, relative to the historian's translation manual, etc. the Greeks believed that the numbers are infinite in the first but not in the second of the two senses I suggested.)

4. POSITING AND KNOWLEDGE

Positing mathematical objects involves nothing more mysterious than the ability to write novels, invent myths, or theorize about

[7] Quine (1969c), 145–6.

unobservable influences on the observable world. For to posit mathematical objects is simply to introduce discourse about them and to affirm their existence.

Yet the very ease with which we can posit generates another worry. People have posited ghosts, the Ether, and phlogiston as effortlessly as they have posited numbers. Why did positing lead to knowledge in the one case and not in the others? One reason is that ghosts, the Ether, and phlogiston do not exist, whereas mathematical objects do. (Remember our epistemological task: assuming the existence of mathematical objects, to explain how we can know them.) Thus no matter how justified people might have been in positing the former, their positing could not have led them or anyone else to knowledge.

Of course, neither the existence of the things we posit nor the truth of what we say about them guarantees that our positing will lead us to knowledge, since we may lack the appropriate justification for our true beliefs. Did our ancestors have this sort of justification? In responding we must separate the question of whether they were justified in *positing* mathematical objects from the question of whether they were justified in *believing* in the things they posited. Positing can be done tentatively or decisively. Thus it might have been that the first ancient mathematicians to posit mathematical objects actually expressed serious doubts about their existence, and no one took their postulates as steadfast affirmations. It takes very little to justify this sort of tentative positing. The only major worry is that it be a waste of time.

If my historical speculations are correct, then clearly ancient mathematicians were justified in trying to extend their mathematical theories by tentatively positing mathematical objects. For in developing and studying systems of numerals and concrete diagrams they had already laid the foundations for theorizing about the new objects. Furthermore, they had reason to believe that the new theoretical framework would allow them to simplify, unify, and extend the mathematical principles they had already developed, tested, and applied.

Positing mathematical objects probably produced significant changes in ancient mathematical practice and hastened the arrival of the mathematical method as we now know it. For the nature of the new objects meant that reasoning from postulates governing them would play a much more authoritative role than perceptual verifications. Assuming that he would concur with the facts of the case, Philip Kitcher would deem this transformation a *rational inter-practice transition*, which, as such, would warrant our ancient mathematicians' believing the new postulates and countenancing mathematical objects.[8] Now Kitcher's approach prompts the worry that while it certainly was scientifically rational for mathematicians to *introduce* and *promote* the new practice, it may not have been rational for them to *believe* in the new objects. This is because they may have had ample evidence for the utility of their new theories but little evidence for their truth.

In Chapter 7 I pointed out that when scientists introduce a new theory they usually postulate principles linking the theory to previously accepted methods for obtaining evidence. This has the effect of extending their local conception of evidence, and allowing them to take certain data to bear upon specific hypotheses of the new theory. Thus although the decision to initiate a new theory is based upon considering the benefits to the more global scientific context into which the theory is introduced, most questions raised within the new theory's framework can be adjudicated using its local conception of evidence. The result is that we count the theoretical framework itself as well-supported so long as its local evidence sustains it, and usually we evaluate it from a more global perspective only when it fails in local terms.

We have already supposed that our ancient mathematicians also postulated links between their new theories and their former methods for obtaining evidence. This allowed them, for example, to use physical calculations that previously counted as evidence for claims about token numerals or countings, as evidence for claims about numbers. It thus gave them an extended conception of evidence to invoke in support of their new theories. Furthermore, in time they found further support for their new mathematics through successfully using it as a framework for much science and technology. (Perhaps their mathematics became indispensable to their science. But using indispensability reasoning would have required taking a global perspective on science and mathematics.)

Whether we assess their justification 'internally' by focusing on the evidence available to them or 'externally' by focusing on the evidence available to us, the case is strong for crediting our ancient

[8] Kitcher (1983), 225–6.

mathematicians with knowledge of mathematical objects. If so, it is plausible that mathematical objects entered the human ken through ancient mathematicians positing them. But this still leaves the question of how *we* came to know about mathematical objects.

Of course, we initially acquired our mathematical knowledge from exposure to teachers who stood at the end of a chain of mathematics teachers reaching all the way back to the ancients. In speaking of *exposure* and a *chain of teachers* I intend to leave open the questions of whether interactions between teacher and students preserved information, reference, or knowledge. Quine's work in the philosophy of language and Kuhn's and others in the philosophy of science give one reason to doubt that it preserved either information or reference. And flaws in historical mathematics, such as Euclid's tacit continuity assumptions, suggest that our teachers do not always impart knowledge.

But I don't see the last point as a real concern. Although mathematicians take a fair amount of mathematics on authority, they also rework, reformulate, and re-prove large bodies of previously accepted mathematics. In this case their teachers are not so much a source of the reworked mathematics as they are initiators of a process leading to new knowledge. Thus it is not necessary that the mathematics of our teachers be free of all errors for us to know some mathematics. Yet the justification of our mathematical beliefs probably depends upon some of their mathematics being correct; simply because mathematicians, taken both individually and as a body, do not re-create all the mathematics they use.

My account of our learning process still assumes that the previous generations held beliefs about mathematical objects. Their holding beliefs having a certain content is unproblematic to all but Quineans, who can accommodate this idea by reference to a translation manual as I suggested in the last section. However, anti-realists might bridle at the idea that *anyone* can hold beliefs *about* mathematical objects on the grounds that, if they exist, they cannot stand in the causal relations necessary to establishing and preserving reference. In the end, my response to this worry will amount to Quine's. With him I will countenance no facts of transcendent reference for mathematical terms. However, they do refer relative to taking our language at face value (and a translation manual if they be foreign).[9]

[9] It will turn out that even taking our language at face value can fail to fix facts of

I shall develop my views on reference to mathematical objects further in the next section and in subsequent chapters.

5. POSTULATIONAL EPISTEMOLOGIES AND REALISM

Whether or not we inherited our knowledge of various mathematical objects from previous generations of mathematicians or were simply prompted by listening to or reading them to posit these objects ourselves, positing remains an indispensable tool of contemporary mathematics. For it is the only way we have for introducing mathematical structures that are more complex than those we already recognize. Thus it is essential to establish that a postulational epistemology is compatible with realism.

To begin, let us remember that we do not create mathematical objects through postulating them. Even constructivists would agree that merely postulating that infinitely many twin primes exist is no more potent than wishing that they exist. Instead positing something is a step towards acknowledging or recognizing a thing that exists independently of our positing it.

Although mathematical posits exist independently of our postulating them, they are not independently given to us. We must introduce terms for them and posit them in order to recognize them. Thus one might well wonder what distinguishes mathematical posits epistemically from fictional characters. Since we require neither mathematical nor fictional objects to be physically detectable, couldn't a literary mathematician write some mathematics that one could read as a piece of fiction? Conversely, couldn't a piece of fiction contain some significant mathematics? Now, of course, we can read mathematics as fiction and also find mathematics in fiction. Usually, very obvious differences in style and vocabulary differentiate articles intended as mathematics from essays in fiction, but the existence of mathematical parables, such as Poincaré's illustration of a non-Euclidean world, shows that the matter does not turn on syntax.

co-referentiality, such as whether number words refer to the same objects as some of the terms for sets do. If we use the idea of the human polyglot put forth in Chapter 2, something similar affects 'foreign' terms. It will be true that, say, the Greek word 'arithmos' refers to *arithmoi*, but there need be no fact of the matter as to whether *arithmoi* are numbers.

Nor does it turn on reference. Fiction can be about real—even mathematical—characters and, contrary to its author's intentions, it might even be entirely true. Furthermore, having a recognizably real subject-matter could not be the test for separating mathematics from fiction, since often when someone introduces us to new mathematical objects through positing them they are not recognizably real. Rather, we mark something as a piece of mathematics by treating it and expecting it to be treated differently from how we treat fiction. We require it to meet different standards of accessibility, clarity, precision, rigour, coherence, and thoroughness. Mathematics also plays a different role in our intellectual life from fiction. Unlike fiction it occupies an (apparently) ineliminable place in our scientific endeavours. Thus the rationale for positing new mathematical objects is quite different from that for creating new fictional characters.

We rarely introduce mathematical posits lightly. Ordinarily they should answer to a clear mathematical need, such as allowing us to answer questions that our previous mathematics left undecided or to systematize and extend a body of previous results. We also seek evidence for mathematical posits, which we do not seek for fictional characters. Our postulates are hypotheses that we are prepared to modify or withdraw in the face of evidence that they are inconsistent, have unwanted models, fail to yield the consequences we seek, or poorly fit our broader mathematical and scientific programmes. In short, we use these and the other means discussed in Chapter 8 to assess our new postulates in terms of our (possibly now enhanced) local conception of mathematical evidence.

By contrast, none of these evidential constraints apply to the stipulations through which authors premiss their fictions. They are extraneous to the practice of fiction writing. Not only do some good stories mock science and common sense, but in principle they might violate elementary logic as well. Finally, unlike mathematics, fiction is not indispensable to science, so we need not presuppose its truth in investigating the world.

Despite this, there is important disanalogy between mathematical positing and scientific positing, which some may find disquieting. Scientists often posit to explain previously observed phenomena. And even when non-explanatory considerations prompt physicists to posit a new kind of particle, good physical practice requires them or their colleagues to develop experiments for detecting the particle. I see no mathematical match for this case. When we introduce new mathematical theories, even those proposed with applications in mind, we do not try to detect the mathematical entities we posit.

Of course, the reason we don't is that we know that mathematical objects are undetectable. But this prompts one to ask why our predecessors did not postulate mathematical objects endowed with detectable features. We have already seen why. They posited mathematical objects for a different purpose from that for which they posited elements, forces, or gods. They posited them to enhance their ability to give precise descriptions and to reason about both physical and mathematical phenomena. Endowing them with physical characteristics would have only led to unnecessary complications. We posit mathematical objects for similar reasons today. We also posit to extend mathematical structures that we already recognize, as Tarski did in positing inaccessible cardinals, or to introduce new methods of description and reasoning, as Cantor did in positing sets and transfinite ordinals. Because we do not posit mathematical objects as explanatory principles or causes, we neither need nor expect them to be detectable. This difference between the motivations for positing physical and mathematical objects may be one reason why physicists seem more suspicious of the undetectable interiors of black holes or virtual processes than mathematicians of the natural numbers.

It should now be clear that positing does not fictionalize mathematics or detract from our justification in recognizing mathematical objects or truths about them. However, combining a postulation epistemology with realism brings another problem to the forefront: making sense of the idea that our mathematical postulates are *about* an independent mathematical reality.[10] To appreciate the worry one might have, consider a team of astronomers who posit a galaxy which they have not yet observed. Suppose that they subsequently observe some new galaxy. Obviously their positing it did not bring it into existence—it was already there. But was it the one to which they had been referring? Suppose, for instance, they had posited a galaxy to account for what appeared to be its gravitational effects on other astronomical bodies, and that the newly observed galaxy had no

[10] In correspondence Pieranna Garavaso wondered how to explain how introducing terms for mathematical objects enables us to refer to entities that exist independently of our positing them.

such effects; then it would seem that they would be mistaken in referring to it as the one they had posited.

One way to account for examples like this is to appeal to the causal theory of reference. If, at the time they posited a galaxy, our astronomers stood in the 'appropriate' causal relationship to the one they subsequently observed then it is the one to which they had been referring. As we know, this account cannot work for mathematical objects.

But incompatibility with the causal theory is no reason for rejecting mathematical posits. In fact, thinking about positing suggests a weakness in the causal theory. For while the causal theory may succeed in the example above, it is difficult to see how it could explain how natural scientists succeed in referring to some of the objects they posit. This is because often purely theoretical, indeed mathematical, considerations rather than experimental ones lead them to posit new objects. Dirac posited anti-matter in attempting to make physical sense of the negative square roots in the equations of special relativity. Group-theoretic reasoning led contemporary particle physicists to posit certain bosons. Yet it is hard to think of any causal processes involving anti-matter or these bosons that would 'appropriately' connect them with the physicists who posited them. In addition, physicists also refer to supposedly physical things that can have no physical effects upon us, such as the atoms inside black holes or virtual processes. As physics advances it seems likely to deal with more and more objects of this sort.

Now I think we should expect a theory of reference to address at least the two following questions:

> *The Genesis Question*: How did we come to use a certain term to refer to a given object?

and

> *The Criterial Question*: When does a given term refer to a given object?

I have heard philosophers say that a theory of reference should also explain what reference is or how it is constituted. But I have no clear conception of what doing so would come to beyond answering the genesis and criterial questions. In any case, I will restrict my attention to these two questions here.

The beauty of a causal-historical account is that it answers both questions in the same way. To explain how we came to use a given term to refer to a certain object it would describe a causal-historical chain leading from us back to the object, and it counts the existence of that chain as necessary and sufficient for that term to refer to that object (rather than to some other object or to none at all). However, a good theory might give separate answers to the criterial and genesis questions. The causal theory seems to be so prevalent among realists because we tend to think of theory construction as a matter of first discovering, then naming, and finally describing reality. But often in positing we describe first, and only later obtain evidence of the independent existence of our posits. Positing suggests that answering our two questions may require two theories (or two parts of a theory) rather than a single unified one.

Just as I distinguished between transcendent and immanent theories of truth in Chapter 2, I also distinguish between transcendent and immanent theories of reference. A theory of reference is transcendent when it purports to be an account of reference in any language. The causal-historical theory is transcendent. Its condition for our name 'The Matterhorn' to refer to the mountain above the Swiss town of Zermatt is that an appropriate causal-historical chain lead from us to that mountain's baptizers. Not only can this condition be formulated independently of English, but also the same type of condition determines the references of foreign names, such as 'Firenze', 'Milano', and 'Praha'. Immanent theories of reference apply only to their own languages, and do not aim to answer the genesis and criterial questions using concepts that are applicable to other languages. Thus we should not expect an immanent account of

> t refers in English to x

to generalize to one of

> t refers in L to x (with variable L),

although a transcendent theory would treat the former as a special case of the latter. Transcendent theorists should find nothing to deny in the immanent answers I shall give to our two questions, although they will, of course, find them incomplete.

The expression 'the present capital of North Carolina' refers to Raleigh, for that city is North Carolina's capital. It does not refer to Hillsborough, since Hillsborough has not been the seat of the North Carolina government for almost two centuries. This example illustrates the general form of an immanent answer to the criterial question:

(Sing) For any x, the singular term 't' refers to x just in case $x = t$,

where t is a schematic letter standing for English singular terms. The condition '$x = t$' makes the answer immanent, because putting most foreign singular terms for 't' in (Sing) renders it inapplicable, by virtue of containing an identity outside our language. (Of course, if, as I proposed in Chapter 2, we take our language to be the human polyglot, then this restriction has no practical consequences.)

We can treat English predicate reference similarly. This is the criterion for reference by (one-place) predicates:

(Pred1) For any x, the predicate 'F' refers to x just in case x is F,

where F is a schematic letter standing for one-place English predicates. Again the criterion is immanent, because substituting most foreign predicates for F will turn part of (Pred1) into non-English. (Obviously, we must exclude indexicals and equivocal terms or else greatly complicate this account. Fortunately, this restriction rules out few mathematical terms, although a number of symbols are equivocal. For example, '+' and '0' have different meanings in Boolean Algebra and Number Theory.)

Using this theory of reference, there is no special problem with referring to mathematical objects. The predicate 'number', for instance, refers to an object if and only if it is a number. End of story. Taking our language at face value, we can affirm truly within it that 'number' refers to numbers and not to mountains or gold bricks. For numbers exist, and, given their abstractness, they are neither mountains nor gold bricks. Furthermore, in using 'number' to refer, we refer to something existing independently of our constructions, proofs, and so on, since our constructing a mathematical object or proving theorems about it is not necessary for its existence.

This immanent (and disquotational) approach to reference does not undercut the independence of the objects to which we refer any more than the disquotational approach to truth does. Just as the truth of the sentence 'there are infinitely many primes' turns upon whether or not there are infinitely many primes (and not upon our so proving), so does the reference of 'the number of apostles' depend upon how many apostles there were (and not upon our opinion of how many).

We began with a worry about making sense of the intuition that astronomers could mistakenly refer to a newly observed galaxy as the one they posited. We can generate a similar intuition concerning a mathematical case. Suppose that some mathematicians introduce the term 'finor' to refer to some new entities that are supposed to be simply ordered and it turns out that the axioms they subsequently introduce for characterizing finors do not imply that they are simply ordered. (Such a simple mistake is unlikely, but remember that Frege thought that a defining feature of extensions was to be associated in a one–one fashion with concepts.) Then just as our astronomers would be mistaken in referring to the galaxy without the posited gravitational effects as the one they had posited, our mathematicians would be mistaken in referring to the structures modelling their axioms as finors. We can handle this example without invoking the causal theory, because 'finor' refers to a model of the axioms just in case its elements are finors.

Notice, however, that our criterion for a term's referring to something answers few questions about its reference, including whether the mathematicians were mistaken in calling the entities introduced by their axioms 'finors'. The answer turned upon whether the models were simple orderings. Nor will our criterion create facts where there are none: if there is no fact of the matter whether 2 is the class of all pairs, then there is no fact of the matter whether '2' refers to the class of all pairs. But this does not undercut our immanent approach to reference, because transcendent theories, and the causal-historical theory in particular, offer no advantage here either. Indeed, the lack of a fact of the matter in this case is due to the incompleteness of mathematical objects. This is a datum we want eventually to explain rather than to abolish.

The other question we want our theory of reference to answer is: how did we come to refer to mathematical objects? On the immanent approach taken here, our standard mathematical terms refer, if at all, to the objects to which our standard mathematical theories are committed. Thus the question comes to asking how we came to use such terms in a referential way, and answering this is a matter of tracing the history of our use of the relevant terms.

Of course, using terms in a referential way is no guarantee that they refer to anything. But, recalling that we have been assuming throughout this chapter that the standard mathematical objects do exist, we know that our standard mathematical terms do refer. This assumption and our immanent approach to reference are crucial to this knowledge, for without them it would seem an inexplicable coincidence that our mathematical terms happen to refer to a mathematical reality that exists independently of our positing it. Given these assumptions, however, it is trivial that 'number' refers to numbers, 'set' to sets, and so on. It is also comparatively easy in this context to explain their independent existence. For contemporary mathematics, together with some obvious premisses about us, imply that numbers and sets and truths about them outrun our current methods of construction and proof. The thought that something deeper than this is at work results, I think, from not tacitly conceiving of reference in transcendent terms.

One might also wonder why positing mathematical objects has been as successful (or reliable) as it has been in generating mathematical knowledge.[11] In approaching this question it is useful to compare mathematical positing with scientific positing. Let us consider first how we know that each has been successful. In the case of science it would be natural to reply that we know that positing has been successful here because scientists have experimentally detected many of the entities they have initially posited. But notice that these detections depend upon scientists endowing their posits with physical properties and using their theories to forge connections between their posits and detectable phenomena. Scientists have been successful in positing in part because they have been able to design their theories so that they can connect the posited objects with data that count as detecting the posits. Of course, mathematicians don't even try to detect the objects they posit, since they endow them with no properties that might be detectable. But they do recognize evidence that the theory of the posits is consistent and bears fruitful connections to other mathematical and physical theories as counting in favour of the existence of the posits. In both cases, then, successful positing is measured in terms of the success of some theory of the posits, where success is ordinarily measured by criteria that are local to the discipline. What we want to know, then, is why mathematicians have been so successful in designing theories that count as successful

[11] Mark Balaguer put this question to me in correspondence.

by the criteria which the discipline of mathematics recognizes. Finding out why this is so may take some work, but it will certainly not require us to make contact with an undetectable mathematical realm or even take a perspective which transcends our current scientific theories. I will not carry this project further here. My purpose has only been to show that the project of explaining the success of mathematical positing can be put into a feasible form.

A final matter that might come to mind in thinking about beliefs about mathematical objects is the distinction between *de dicto* and *de re* beliefs. Now this distinction concerns the ways in which one has grasped or is related to or is in epistemic contact with an object. Mathematicians might believe, for example, that there is a final pair of twin primes without believing of any pair of numbers that they are this pair. (Perhaps their evidence is non-constructive.) In this case, their belief is supposed to be a *de dicto*, but not *de re*, belief about twin primes. By contrast someone who has calculated that six is the smallest perfect number presumably has a *de re* belief about six. Notice that this distinction is predicated upon our already having the ability to form beliefs *about* the entities in question. Thus while it may well be worth further investigation, it is not one that we need treat in explaining how we acquired knowledge about mathematical objects.[12]

[12] In Resnik (1990a) I argued that whether we attribute a *de re* or *de dicto* belief about a mathematical object to someone varies with the context in question and our interests in describing it.

PART THREE

Mathematics as a Science of Patterns

INTRODUCTION TO PART THREE

In Part Two I presented as much of an epistemology for mathematics as I thought I could without assuming anything about the nature of mathematical objects except for their abstractness. One reason that I got as far as I did is that mathematics does not say anything about the nature of mathematical objects in general. (Of course, it says plenty about numbers and other specific kinds of mathematical objects.) However, before my account can be complete, I must face some important tasks. Ironically, one of these is to explain why mathematics doesn't say anything about the general nature of its objects. We saw (Chapter 5) that this is related to the incompleteness of mathematical objects. So one of my aims in this part will be to come to grips with this phenomenon.

In the last chapter I claimed that by positing links between paper-and-pencil computations and diagrams and certain mathematical objects, we can use things that are physically accessible to gain knowledge about abstract mathematical objects. For example, by assuming that a sequence of unary numerals shares structural features with an initial segment of the natural number sequence, we can learn features of the latter from those of the former. But I coasted over the question of why it would be plausible to assume that the numbers and numerals share structural features and what those features might be. I did not explain why, for instance, numbers do not have shapes although numerals do. I mean to fill this gap in my epistemological account here.

Below I will present a view of mathematical objects capable of completing these tasks. It is the view that mathematics is a science of patterns with mathematical objects being positions in patterns. In Chapter 10 I explain the notions of pattern and position, and relate them to the incompleteness of mathematical objects, reduction in mathematics and reference to mathematical objects. In Chapter 11 I return to epistemological questions and show how to connect mathematical knowledge with the study of patterns. In Chapter 12, which deals with questions and objections that the pattern view raises, I attempt to show that despite its controversial features the view does not substitute a new set of equally intractable problems for those it claims to solve.

10

Mathematical Objects as Positions in Patterns

1. INTRODUCTION

Much philosophical thinking about mathematics is guided by a fun-mental misconception of the subject-matter of mathematics. We tend to conceive of the numbers, say, along the same lines as we think of tables or stars or even atoms. That is, we think of them as objects that can be discussed and known in isolation from the others. This is why we are likely to think that discovering something about a number should depend upon some sort of interaction between us and that number, and to believe that this ought to be reflected in a satisfactory epistemology for mathematics. It is also why we are likely to hold that the identity of a number *vis-à-vis* any other object should be completely determined. However, for some time the practice of pure mathematics has reflected the idea that mathematics is concerned with structures involving mathematical objects and not with the 'internal' nature of the objects themselves. Mathematicians as prominent as Dedekind, Hilbert, and Poincaré have even voiced structuralist ideas in reflecting philosophically on mathematics.[1] More recently the structuralist approach has been credited with fostering category theory.

The underlying philosophical idea here is that in mathematics the primary subject-matter is not the individual mathematical objects but rather the structures in which they are arranged. The objects of mathematics, that is, the entities which our mathematical constants and quantifiers denote, are themselves atoms, structureless points, or positions in structures. And as such they have no identity or distinguishing features outside a structure.[2]

[1] See Dedekind (1963), Hilbert (1971), Poincaré (1913), 43. See Parsons (1990) for additional historical information.

[2] A bit of history: during the last thirty years philosophers began to note the

For epistemological purposes I find it more suggestive to speak of mathematical patterns and their positions rather than of structures. For this brings out important similarities between mathematical knowledge and other knowledge such as grammatical and musical knowledge. But the term 'structuralism' has a better ring to it than 'patternism', so the philosophical literature invariably speaks of structures instead of patterns. In what follows I will use the terms 'pattern' and 'structure' more or less interchangeably.

As we will see, most mathematical theories do not refer to the patterns they attempt to describe, and thus do not treat them as entities at all. Were one to develop a mathematical theory in which patterns or structures figured as mathematical objects, one would treat them as positions in a pattern of patterns. In Chapter 12 I will say more about when and whether to consider patterns as entities. For now I will simply try to explain my conception of patterns.

2. PATTERNS AND THEIR RELATIONSHIPS

I know of no developed philosophical account of patterns. The suggestions I have encountered in conversation and correspondence tend to characterize a pattern as either a kind of universal or an equivalence class of instances, and they begin with the relationship between a pattern and its instances. I will start instead with relationships between patterns and get instantiation as a special case. This leads to an extensional view of patterns, whereas thinking of them as universals might lead one to a more intensional theory. The approach seems quite natural to me, but my experience has been that people's intuitions vary considerably on these matters. It may help you understand my approach if you bear in mind that I arrived at it by reflecting on geometry and model theory, the two mathematical theories of structure to which I have been most exposed. Metaphorically, my view makes model theory into a geometry.

I take a pattern to consist of one or more objects, which I call

structuralist movement in mathematical thought, and somewhat tentatively began trying out structuralist approaches to the philosophy of mathematics. (Cf. Jubien (1977), Kitcher (1978), Parsons (1965), Benacerraf (1965), Steiner (1975).) Beginning in Resnik (1975) I began to develop an explicit structuralist approach to the epistemology and ontology of mathematics, which I articulated in Resnik (1981).

positions that stand in various relationships. For generality I will allow that some of these relations may be monadic and that some of the positions may be 'distinguished'. This permits the colours of the positions in a visual pattern or the specific notes (or sounds) in a musical pattern to be as much a part of the pattern as their arrangement. Although mathematicians often consider patterns with distinguished positions, for example, the natural numbers under successor with zero as distinguished position (*N*,*S*,0), their theories can usually be reformulated to treat another pattern in which no position is distinguished. Thus one can use (*N*,*S*) instead of (*N*,*S*,0) because 0 can be defined as that position in (*N*,*S*) which is not a successor.[3]

A position is like a geometrical point. It has no distinguishing features other than those it has in virtue of being the particular position it is in the pattern to which it belongs. Thus relative to the equilateral triangle *ABC* the three points *A*, *B*, *C* can be differentiated, but when considered in isolation they are indistinguishable from each other and any other points. Indeed, considered as an isolated triangle, *ABC* cannot be differentiated from any other equilateral triangle. Geometry reflects this by focusing on structural relationships, such as congruence and similarity, and reserving claims about the identity of geometric objects for contexts where they are related to other geometrical objects. I transfer this geometric analogy to the various structures studied by mathematics. Within a structure or pattern, positions may be identified or distinguished, since the structure or pattern containing them provides a context for so doing. However, just as in geometry, the premier relationships among patterns are structural ones, namely, structural similarity (pattern congruence and equivalence) and structural containment (pattern occurrence and sub-pattern).

Here is how this geometric analogy applies to the natural number sequence (*N*,*S*). I take this to be a pattern with a single binary relation (successor) and the natural numbers to be its positions. Viewed this way, there is no more puzzle to the natural numbers lacking identifying features beyond those definable in terms of this pattern than there is to the corresponding fact about the points in triangle *ABC*.

[3] This sort of trick will not work for (Int,*S*,0), i.e. the integers under successor with 0 distinguished. Describing (Int,*S*,0) either requires taking 0 (or something interdefinable with it) as a primitive or else defining (Int,*S*,0) in terms of some other structure, such as (*N*,*S*), in which a distinguished position can be defined.

Patterns are related to each other by being pattern-congruent or structurally isomorphic. In Hilbert's formulation of geometry there are different congruence relationships for line segments and angles. Other branches of mathematics must distinguish between different kinds of isomorphisms (group isomorphisms, ring isomorphisms, etc.) so as not to confuse structures that are isomorphic as, say, groups but not as rings. I similarly recognize different pattern-congruence relations for different types of patterns. But just as informal mathematical talk leaves the type of congruence or isomorphism understood, I will do so here. Also for brevity I will simply use the term 'congruence' rather than 'pattern-congruence' where the context is sufficient to distinguish it from geometric congruence.

Pattern-congruence is an equivalence relation whose field I take to include both abstract mathematical structures and arrangements of more concrete objects. Thinking of patterns as models of formal systems, it is the relationship which holds between isomorphic models of formal systems. Consider, for example, a first-order system S with a single two-place predicate 'R' and axioms stating that R is a total ordering. This system has many models—all the total orderings—but they are not all congruent to each other. Only those whose domains have the same cardinality are. The set of numbers from one to ten taken in their natural order and ten pennies stacked on each other taken in order from top to bottom are isomorphic models of the system S; so I count the abstract numerical structure and the stack of pennies as congruent.

When a pattern and an arrangement of so-called concrete objects such as the pennies are congruent then I say that the arrangement *instantiates* the pattern. Instantiation then is a special case of congruence in which the objects 'occupying the positions' of a pattern have identifying features over and above those conferred by the arrangements to which they belong. The pennies thus instantiate the one-to-ten pattern.

One reason that I have not defined patterns via their instances, say, as isomorphism types or classes of arrangements, is that this would require a prior ontology to instantiate mathematical patterns. Otherwise, all universal quantifications concerning uninstantiated patterns would be vacuously true. Science and common sense do not recognize enough non-mathematical things to guarantee that all mathematical patterns are instantiated. So those defining patterns via their instances must posit things (presumably mathematical things) to instantiate the more extensive and complex mathematical patterns. (It is arguable that these would include all infinite patterns.) But positing *mathematical objects* that are not themselves taken as positions in a pattern is to give up a basic structuralist thesis.[4] Moreover, the approach I have taken seems to follow mathematical practice: when mathematicians cannot model a structure within one they already accept, they posit additional mathematical objects as elements of the structure in question without attempting to say more about their nature.[5] In my terminology, they posit positions related as the contemplated structure requires.

Another way in which patterns may be related is that one may occur with another. To understand this, consider the linguistic pattern or sentence, 'A cat bit the cat who ate the cat.' The word 'cat' is also a pattern of letters which occurs three times in the sentence. By analogy, I take the natural number sequence to occur within real numbers taken in their natural order, within the iterative hierarchy of sets, and within itself. Five pointed solid star patterns occur in both the patterns of the United States flag and the California flag. Just as patterns can occur within other patterns, concrete arrangements can occur within concrete arrangements. But my interest is mainly in patterns occurring within patterns. Thus I will characterize *occurrence* as a reflexive and transitive relation which holds between structures P and Q when P is isomorphic to a structure R whose positions are those of Q and whose relations are definable in Q. Thus the structure (N,S) occurs within $(N,<)$ and within models of various set theories. The rational numbers *qua* field (Rat,+,x) contain an occurrence of (N,S), but the rational numbers *qua* countable dense ordering (Rat,<) do not, because 0 and successor are not definable in such structures. On the other hand, although the real numbers and their ordering are set-theoretically definable from the natural numbers, (Real,<) does not occur in (N,S) because the latter does not have enough positions.[6]

[4] One way to obtain mathematical objects instantiating infinite mathematical structures is to take a non-structuralist approach to set theory, and use it to construct models of other mathematical theories.

[5] There are many examples of this, but the introduction of the complex numbers to obtain an algebraic closure of the reals illustrates what I have in mind.

[6] I have not specified the notion of definability to be used in this definition. I will say more about this below. Marcus Giaquinto showed me how to improve this definition over the one I gave in Resnik (1981).

A special case of pattern occurrence is the sub-pattern relation. A pattern *P* is a *sub-pattern* of another pattern *Q* just in case *P* occurs within *Q* and every position of *P* is a position of *Q*. It follows that a pattern occurs in another just in case it is congruent to a sub-pattern of the latter.

The sub-pattern relation is more complicated than the sub-model relation of model theory, although it includes an analogue to that relation in its special cases. There are two reasons for taking this more complex approach: (1) it is necessary to capture certain intuitions about patterns, and (2) it provides a nice account of reduction in mathematics.

I will take up the intuitive considerations now and leave the discussion of reduction for later. Suppose we consider a pattern that may be described as having nine distinct positions a, b, c, d, e, f, g, h, i, and two relations A and L such that aLb, bLc, dLe, eLf, gLh, hLi, aAd, dAg, bAe, eAh, cAf, fAi. To get a better grasp of the pattern consider the following diagram, where L corresponds to the relation *one unit to the left of* and A corresponds to the relation *one unit above*.

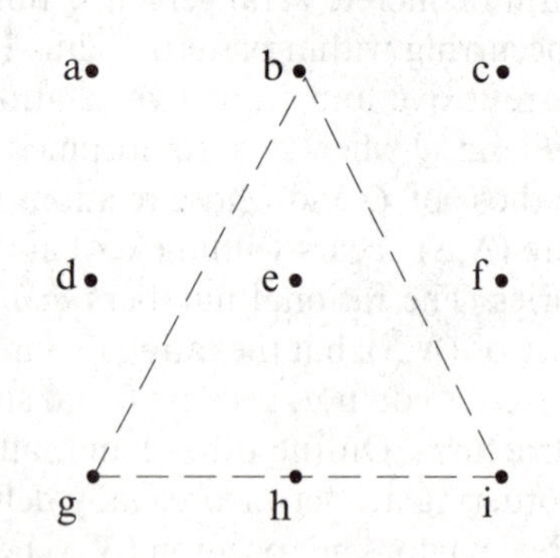

Now as I look at the diagram (and I admit that you may not see things as I do) I take the rectangle of dots to contain a triangle of dots gbi (and other dot triangles as well, of course). Thinking now of my drawing as also a representation of an abstract, spatial dot pattern (which is not the original L–A pattern, but a more complicated one), I conclude that the spatial dot triangle is a sub-pattern of the whole spatial dot array. Then analogical thinking leads me to assume that corresponding to the dot triangle there must be a sub-pattern of the original pattern involving the positions g, b, and i. This pattern can be characterized by relations definable in terms of A and L, just as the spatial dot triangle can be described in terms of relations definable in terms of *above* and *left*. (The triangle consist of the dots g,b, i where g is related to b by being two units below and one to the left of it, etc.) Indeed the former definitions can be read off the latter quite straightforwardly. Nevertheless, the relations of neither the spatial nor the non-spatial sub-pattern are simple restrictions of the relations of the original pattern. Thus in order to count these as sub-patterns of the larger patterns the sub-pattern relation must be defined along the lines which I have suggested.

My visual intuitions with respect to sub-patterns and pattern occurrences can be transferred to some of the model-theoretic cases as well. Let us think of the natural number sequence as represented by means of an unending linear sequence of dots:

• (and so on).

Clearly the sequences of dots which we obtain from this sequence by starting with the *n*th dot from the left are sub-patterns of the original pattern. These correspond to all those progressions which are obtainable from (N,S) by restricting S to a subset of N. But there are infinitely many models of number theory obtainable from (N,S) which are left out because their relations of succession are not sub-relations of the successor relation. The even number sequence is one, the odd number sequence is another, and the prime number sequence a third. These all correspond to selecting some progression of dots from the original sequence of dots. I see these sequences as occurring within the original sequence, and to account for this both the sub-pattern and pattern occurrence relations must be characterized in terms of definability.

The natural number sequence (N,S) occurs within the natural numbers ordered under less than, $(N,<)$, since:

$$Sxy \leftrightarrow x < y \;\&\; \neg(\exists z)(x < z \;\&\; z < y);$$

and, allowing second-order or set-theoretic definability, the converse relation holds too, since:

$$x \leq y \leftrightarrow (\forall z)(x \in z \;\&\; (\forall u)(\forall v)(u \in z \;\&\; Suv \rightarrow v \in z) \rightarrow y \in z).$$

Most mathematicians and logicians would regard number theory developed in a language in which the successor symbol is primitive as essentially the same as a version taking the symbol for *less than* as primitive. Since I am viewing number theory as the science of a certain pattern or patterns, this would suggest that (N,S) and $(N,<)$ should count as the same or essentially the same pattern. The problem then is to find a satisfactory characterization of this relation. Pattern congruence is too strict a condition, since these two structures are not isomorphic. Moreover, they are not isolated examples: mathematicians have defined Boolean algebras as a kind of ring, but also as a type of lattice, have given alternative definitions of groups and topologies, and so on.

It happens that each of these examples concern theories whose patterns occur within each other. However, mutual pattern occurrence seems to be too weak a condition for pattern equivalence. One reason is that $(N,S,+,x)$ and $(\mathrm{Rat},+,x)$ occur within each other, although they are not regarded as essentially the same. (Fractions can be coded as ordered pairs (m,n), and these as numbers of the form $2^m 3^n$.) A second reason is that the pattern-occurrence relation corresponds to the notion of the interpretability of theories (by means of definitions) and there are theories which are mutually interpretable but not essentially the same from the metamathematical point of view.[7]

If we look at well-known examples of 'essentially the same' theories we see that they meet a stronger condition than that of mutual interpretability. They are *definitionally equivalent*. This means that there are two sets of definitions Def-S-in-T and Def-T-in-S, such that S + Def-T-in-S yields both T and Def-S-in-T, and also T + Def-S-in-T yields both S and Def-T-in-S. In other words, the theories together with the interpreting definitions yield not only each other but also each other's interpreting definitions. Think of definitions as a kind of axiom and think of the language of a theory as including the symbols introduced by definition. Then definitionally equivalent theories are equivalent axiomatizations of the same set of theorems.

To capture the notion of patterns which are essentially the same we need a relationship like that of definitional equivalence. The

[7] A theory S is interpretable in a theory T with the same underlying logic just in case there is a set, Def-S-in-T, of definitions of the primitives of S in T, which when added to T yields the theorems of S (as theorems of T + Def-S-in-T). See Corcoran (1980) for examples of mutually interpretable theories that are not 'essentially the same' theories, and for further discussion. I am indebted to John Corcoran for this point.

thought that definitionally equivalent theories are distinct encapsulations of a common set of theorems suggests a similar approach for patterns. The idea is to take two patterns as 'essentially the same' if they both encapsulate some 'bigger' pattern from which they can be obtained by deleting some of its relationships. More formally, let us call a pattern P a truncation of a pattern Q if every position and relation of P is also one of Q. Then a pattern P will be said to be equivalent to a pattern Q just in case there is a pattern R which is a sub-pattern of both P and Q and of which both P and Q are respective truncations. To illustrate this concept, return to (N,S) and $(N,<)$. These are truncations of $(N,S,<)$ and it occurs within both of them. So they are equivalent according to the definition. On the other hand, although (N,S) and $(\mathrm{Rat},+,x)$ occur within each other they cannot be obtained from a common extension by means of truncation since cutting a progression (N,S) from an open dense ordering $(\mathrm{Rat},S,+,x)$ requires deleting positions from the latter.

3. PATTERNS AND POSITIONS: ENTITY AND IDENTITY

Of the equivalence relationships which occur between patterns, congruence is the strongest, equivalence the next, and mutual occurrence the weakest. Yet none of these is suitable as an identity condition for patterns, because, presumably, identical patterns have the same positions whereas equivalent, mutually occurring, and congruent patterns need not. Considering the strongest of these relations, congruence, suffices to illustrate the problem. A pattern can be congruent to a sub-pattern which does not contain all its positions. For instance, the natural number sequence is congruent to the sequence of even natural numbers. But if we took congruent patterns to be identical, then we would be forced to count the even numbers as the same as the natural numbers. Of course, we might avoid this by restricting substitutivity of identity, as one does in intensional contexts, but that hardly seems the appropriate thing to do here.

This suggests that we identify patterns just in case they have the same positions and the same relations, with the latter being given the usual extensional identity conditions. And I would do exactly this were I to speak of patterns as identical. But I hesitate to speak of them in this way. For unless we make radical revisions in our logical notation, speaking of patterns as identical or distinct is to treat them as individuals, since identity is a relation between individuals. But, as we will see (Chapter 12), there are reasons to avoid being forced into treating patterns as individuals or even as entities of any kind. Furthermore, the proposed identity-conditions seem to require quantifiers ranging over a universe composed of positions from all patterns. This in turn would presuppose facts concerning identities between these positions. Yet there are also good reasons for not assuming that there are facts of the matter as to whether positions from non-overlapping patterns are identical.

Let us turn to these reasons. Recall that positions are like geometrical points in having no identifying features beyond those they acquire by being in a pattern. Now suppose that we are told that a point *A* is the corner of a right triangle and the point *A′* is the corner of a rectangle. Are *A* and *A′* the same? One might respond that they cannot be because they belong to separate figures. But we have not been told that they are separate. It might be that the corner of the triangle is a corner of the rectangle. Of course, this is no reason to conclude that there is no fact of the matter here concerning the identity of *A* and *A′*. They are, we may assume, points in the same space and have adequate identity-conditions—such as being on the same lines. We simply do not have enough information to determine the facts in this example.

While this is quite true, its being so is a consequence of developing geometry as a theory of space. Had geometry developed instead as a theory or collection of theories of figures or shapes, then points might only play their role of marking locations *within* figures without marking locations in a containing space. Then we would have regarded the triangle and rectangle as either separate designs or a single composite one, and would take a similar view of other of simple and compound geometrical figures. For our geometry would attribute no being to points independently of the figures containing them. We also would have restricted identity to elements of the same figure, or else have introduced for each type of figure a separate theory with an identity predicate ranging over all elements of its universe.

The latter alternative is the one we find in mathematics. Number theory, for example, is intended to deal with a certain structure; it has the means to raise and answer questions concerning the identity of various numbers, but it cannot even formulate the question as to whether the number one is the real number *e*. Yet within number theory identity is absolute, for 'any numbers *m*,*n* are identical or distinct' is one of its theorems. Similar remarks hold for the theory of real numbers and set theory. Each theory was developed to speak only of elements of a certain structure and has no means to identify or distinguish these from elements of another structure. Like the variant of geometry using different theories for different shapes, mathematics is largely a conglomeration of theories each dealing with its own structure or pattern and each forgoing identities leading outside its pattern.

This mathematical practice has resulted in the so-called incompleteness of mathematical objects (see Chapter 3). But what would strike one as a problematic oversight if one thought of mathematical objects along the lines of ordinary objects, seems quite natural when it comes to positions in patterns. For restricting identity to positions in the same pattern goes hand in hand with their failure to have any identifying features independently of a pattern.

The typical mathematical theory also excludes the structure it is supposed to describe from its universe of discourse—it does not recognize it as a mathematical object. Thus number theory quantifies over the numbers but not over the number-theoretic structure; set theory quantifies over sets but not over the set-theoretic hierarchy. Even model theory, which purports to treat of arbitrary mathematical structures, does so by positionalizing them, that is, by identifying structures with sets or ordered *n*-tuples. Nor does number theory have the means to raise questions concerning the identity of the natural number sequence *qua* structure. The same is true of set theory and the iterative hierarchy. Treating patterns as individuals would undo these parallels with mathematics, since it would allow identities between patterns, and this in turn would permit identities between their positions.

However, not recognizing patterns as individuals and restricting identity to positions within the same pattern threatens to undermine the rest of the theory of patterns I have been developing during the course of this chapter. For a formal version of that theory would seem to treat patterns as entities by using quantifiers that range over them, and countenancing isomorphisms between patterns seems to require a common universe of positions. This is a serious worry. I will begin to address it at the end of Section 5 below, and will continue to discuss it in Chapter 12. To prepare the way for that discussion, I now want to sketch some alternative ways in which one might develop a formal theory of patterns.

Speaking of relationships among various patterns requires a comprehensive theory in which positions from different patterns can be dealt with from a single vantage-point. One option for such a theory uses a many-sorted logic with separate universes for the positions of each pattern and unreduced functions between these universes to serve as isomorphisms. It also requires a universe for patterns, and, as is usual in many-sorted theories, prohibits trans-universe identities. By postulating that patterns are identical only if they have the same positions, it avoids the unwelcome result of one pattern being identical to another without there being a fact of the matter as to whether their positions are. One problem with this approach is that it requires uncountably many different styles of variables, since there are (presumably) uncountably many patterns. Thus its language would be highly infinistic.

A second (logically cleaner) approach constructs a pattern theory, along the lines of geometry, by positing a space of positions from which patterns could be 'constructed' as *sui generis* entities. This theory recognizes facts of the matter for identities involving positions from the different patterns it countenances, since identity would apply to all positions in its space. Two sorts of variables can be used to reflect the distinction between patterns and positions. This theory could also follow the treatment of points used in geometry, and prove identities only between those positions given as belonging to the same pattern. A third approach refines the second by reducing patterns to sets of positions just as model theory construes models as sets built up from the domain of the model. One could push this even further and construe positions themselves as pure sets, thereby reducing so-called pattern theory to set theory.

In a sense these formal theories of patterns, through having positions that represent both positions and patterns that previously fell outside the scope of identity, contravene my earlier claims about identity. But in another sense they do not. For I have not claimed that the various patterns mathematics studies (the natural number sequence, the real line, etc.) are entities or *identical* to the patterns treated as positions within one of the comprehensive accounts, but rather that they could be reduced to them. To appreciate the difference we must turn to the subject of reduction in mathematics.

4. COMPOSITE AND UNIFIED MATHEMATICAL OBJECTS

Before taking up mathematical reductions, I want to respond to a common objection to structuralist accounts of mathematics. The objection is that structuralists cannot account for the fact that certain mathematical objects, such as sets, vectors, and spaces, are composed of others, because structuralists focus on relations between mathematical objects and not on the objects themselves. A related objection is that structuralism fails to explain why certain mathematical patterns are unified wholes while others are not. For instance, some think that an ontological account of mathematics must explain why a triangle is not simply a 'random' set of points.

In responding I will grant that it is certainly true that we tend to think of sets, vectors, triangles, and spaces as made up of other mathematical objects. It is also true that certain arrangements of points (or other collections of mathematical objects) strike us as having unity while others appear entirely random. These features of mathematical experience should not be neglected, as they probably provide the keys to explaining why mathematics developed as it did.[8] But I deny that they are objective features of mathematical objects or structures, that obtain independently of the way we think of these structures and objects. Take sets, the paradigmatic compositional mathematical objects. The mathematics of set theory applies to any iterative hierarchy whether it is generated by a compositional relation or not. That is why, from the point of view of set theory, sets are simply positions in iterative hierarchies. They appear to us to have an internal structure only because we use compositional language and analogies in elucidating their relationships. But while this heuristic may be essential for our thinking about the hierarchy, and perhaps even in grasping axioms and proofs, it is not reflected in the content of set theory itself. For otherwise it could distinguish compositional iterative hierarchies—those with a 'real' membership relation—from those that have no 'real' membership relation.

In replying to the second objection, let me begin by remarking that, by virtue of its generality and abstractness, mathematics

[8] See Grosholz (1991) both for the source of the objections of the last paragraph and for a valuable discussion of the importance of treating certain mathematical objects as unified wholes.

commits itself to geometric objects, functions, and sets whose features strike us as 'weird', 'pathological', or without unity, and even as not worthy of mathematical study. But again this is simply a reflection of our interests and values, and not a matter of objective features of mathematical objects. It is an objective feature of certain functions that they cannot be defined by certain kinds of equations, but it is only a reflection of our interests and values that they are 'pathological' or 'disunified' or 'lawless'. Indeed, certain functions, which were once not even considered to be functions at all, eventually were taken to be of great mathematical significance. In fact, the current set-theoretic notion of a function evolved to accommodate totally 'lawless' or disunified functions.[9]

Let me address another worry. It has been suggested to me that while there may be no fact of the matter as to whether the natural number 2 is identical to the Zermelo Two, surely the natural number 2, the positive integer 2, the positive rational 2, the positive real 2, and the positive complex number 2 are all the same. Far from being silent on this issue, mathematical practice seems to mark no distinction between these numbers. For example, there seems nothing amiss in saying that

> equations of the form '$x^2 = a$' have complex roots for any choice of a, but their roots are *integers* only when a is the square of a *natural number*,

which treats the integers and natural numbers as a species of complex number.

I don't think we should make much of this facet of mathematical practice. First, it may be just a manner of speaking. For we can easily paraphrase the last example as asserting that the roots are *integral* only when a has a *non-negative*, *integral* square root. More generally, we can account for mathematical practice by distinguishing the real complex numbers from those that have imaginary parts, the rational, real complex numbers from the irrational ones, and so on, which is to recognize divisions within the complex numbers without thereby identifying the reals, rationals, or integers with these complex numbers. Second, mathematicians do not speak with one voice on the issue of whether these numbers are the same. For example, when defining the real numbers as sets, sequences, or series

[9] See, e.g. Davis and Hersh (1981).

of rationals, some mathematicians carefully distinguish the rationals from the 'rational' reals (that is, those generated by a rational). Others simply define the reals implicitly as any system of things that satisfies, say, the axioms for a continuously ordered field.

For the most part, however, mathematicians take the numbers as given, and don't even consider the question as to whether, say, the rational number 2 is identical to the real number 2. I suspect that if asked whether they are, their first answer would be 'Sure. What else could they be?' But such a reply is based upon the natural presumption there is a fact of the matter here. This presumption is not a piece of mathematics but rather a piece of background common sense. And as such it is open to revision in light of deeper considerations.

Notwithstanding these points, we can make sense of the claim that the natural number 2 is the complex number 2 by likening the historical development of the complex number system to the step-by-step construction of a complicated pattern through adding positions to an initially simple one. Think of this in diagrammatic terms. Suppose that one starts with a dot diagram representing positions in a pattern, and adds more dots to it. Then the old dots are part of the new diagram; so the old positions are represented as among the new. Furthermore, if one marked the original dots with a colour, say, then they would not be lost among the new ones. Now until mathematicians started explicating the higher numbers as sets or ordered pairs of lower ones, they tended to think of the integers, rational, real, and complex numbers as additions to the previously accepted number systems. From that point of view, one adds, say, the irrationals to the rationals in constructing the real number system, but one doesn't add 2, −3, or ½, because they are, so to speak, already there. To describe this in terms that I will use in Section 6 below, suppose that we take the term 'the natural numbers' at face value, *and also* describe the complex number system as the extension of the natural numbers obtained through closing the latter under algebraic operations. Then not only will there be a fact of the matter as to whether the natural numbers are among the complex numbers, it will be true that they are.

5. MATHEMATICAL REDUCTIONS

The natural number sequence, *qua* pattern, has multiple occurrences within iterative set-theoretic hierarchies, *qua* patterns. Consequently even categorical versions of number theory can be interpreted as dealing with any of these patterns and the set of its true sentences will remain intact. But there is no one correct interpretation. Reflecting on geometric cases explains why this is as it should be. If we completely describe the relationships of a set of points in a geometrical figure, that description continues to hold if we think of that figure as embedded in a larger one. The difference is that our description will not cover the relationships these points have to new points of the containing figure and will fail to characterize that figure.

Philosophers have called developments of number theory within set theory *reductions* of number theory to set theory. Generally when the pattern or patterns one theory treats occur within the pattern or patterns another theory treats, the former can be reduced to the latter. Now when one is interested in proving that one theory is consistent relative to another or in comparing the expressive and deductive strength of theories, it can be important to know that one theory can be reduced to another. In practice, however, the reduction of number theory to set theory has had little effect on mathematics at large. Most books on set theory develop the natural numbers in terms of sets, because doing so illustrates the power of set theory and leads to more elegant formulations of certain theorems. For example, using von Neumann's reduction allows one to derive both finite and transfinite induction from the well-foundedness of set membership. On the other hand, books on number theory, topology, algebra, or analysis often begin with both numbers and sets.

The reduction of the real numbers to sets, converging sequences, or infinite series of rational numbers had considerable mathematical significance when Dedekind, Weierstrass, and Cantor presented these results, because they allowed mathematicians to avoid basing real analysis on questionable geometric intuitions. It also provided standard, rigorous methods for specifying specific real numbers, as one might need to do in proving existence theorems or results about specific reals, such as *e* or *pi*. Frege famously argued that number theory suffered from a similar lack of rigorous foundations. With hindsight we can see that from the mathematical point of view, determining the correct reduction of the real numbers or natural numbers was less important than the non-reductionist projects of characterizing the natural number sequence and real line or axiomatizing the standard theorems of number theory and real arithmetic.

Another phenomenon which has greatly changed mathematics has been the adoption of set theory as the background framework for working mathematics. This has led to treating as sets certain structures, such as the natural number sequence or the real line, which were previously not treated as mathematical objects at all. Put in my terms, it has resulted in representing as positions certain patterns, which were not previously treated as positions. One can see this change by looking at the different emphases of pre-set-theoretic and post-set-theoretic mathematics. Courant and Robbins's famous book *What is Mathematics?* illustrates the pre-set-theoretic approach to mathematics (although it does contain discussions of sets). The book is almost exclusively about the elements of traditional mathematical structures—the various numbers, geometrical objects, and functions. Most of its theorems can be formalized in the first-order theory of these objects. Loosely speaking, these results were obtained by working *within* structures rather than by working *with* structures. Compare this with recent books on topology or algebra. Here the emphasis is upon the various structures themselves, what their substructures are, and how they are related to other structures. To formalize such theorems we must use theories in which structures themselves can be taken as individuals, and this has been greatly facilitated by the introduction of set-theoretic methods.

Category theory also promotes working with structures as opposed to working within them. In the category of vector spaces, for example, the positions are vector spaces and morphisms between them, and the theory of this category attempts to describe how these positions are related. Indeed, under the categorical approach to mathematical objects one tries to show how their supposedly internal features can be characterized solely in terms of their relationships. This is nicely illustrated through the category-theoretic definition of a function f's being one–one (monic). Instead of the usual

(1) f is 1–1 iff $f(x) = f(y)$ only if $x = y$, for all arguments x,y,

category theory uses

(2) f is 1–1 (monic) iff $fg = fh$ only if $g = h$ for all functions g and h,

which transfers the matter from the nature of the function's ordered pairs to the way in which it composes with other functions. Here instead of treating functions as composed of their arguments and values, one treats them as positions in the pattern generated by the composition relation.

Thus post-set-theoretic mathematics is no exception to the rule that the objects of a mathematical theory are the positions in the structures it describes. And even when speaking of structures as mathematical objects, it does so by taking them as elements of 'larger' structures, such as a hierarchy of sets or a category.

We now have two kinds of reduction. With the first, one pattern occurs in another and the theory of the former is reiterated in the theory of the latter. With the second, patterns themselves are *positionalized* by being identified with positions of another pattern, which allows us to obtain results about patterns which were not even previously statable. It is the second sort of reduction which has significantly changed the practice of mathematics. To these two we might add a third kind, which often arises when mathematical modelling is applied to mathematics itself. I have in mind such 'reductions' as those provided in computability theory, where algorithms are variously identified with Turing machines, recursive functions, and so on, and proof theory, where syntax is arithmetized.

With each sort of reduction, we are tempted to ask whether the reduced things are in fact identical to the reducing ones. Philosophers have been fascinated by the first kind of reduction, wanting to know what numbers really are. And we now see that they might also have wondered whether the natural number sequence is a set, or whether all algorithmic functions are recursive, as indeed some mathematicians have.

These questions are ambiguous. Taken one way, the question about numbers, for instance, simply asks whether certain sets are sets. To be interesting, the questions should be taken as asking whether certain things known to us before the advent of set theory (or recursion theory, etc.) are sets (recursive functions, etc.). Since, on my view, sets (recursive functions, etc.) are positions in certain patterns, I take the questions to ask whether things which are not known to be or not given as positions of certain patterns are positions in them. The very fact that our language allows us to invent singular terms more or less at will and to form identities between them gives these questions prima facie sense. But how could we find their answers? Not by mathematical means. No new theorems will settle them. Taken non-trivially, these questions are not even in the language of a single mathematical theory. They can, to be sure, be 'resolved' by forming the union of the relevant languages and adding new identities (or non-identities) as axioms. However, (local) mathematical evidence cannot decide between various resolutions, since it makes no difference to the content of mathematics whether, for instance, we do set theory with number theory on top of it, so to speak, or whether we do it with number theory reduced to it.

Thus mathematical reductions cannot tell us what numbers, functions, and sets are in any factual sense, but they can lead us from a defective conception or theory of mathematical objects of a given kind to more adequate ones. For example, to someone who objects to functions on the grounds that they involve intensional notions such as rules, we might respond that functions can be treated as nothing but sets of ordered pairs. Or to someone who thinks that the foundations of mathematical analysis are actually geometric intuitions of limited reliability,[10] we can reply by pointing to the arithmetical foundations of analysis. There is nothing wrong with this sort of response, so long as we realize that we are not claiming that functions conceived as rules are the same as functions conceived as sets or that a geometric curve is identical with a set of numbers. What we are doing is pointing out that we can discard worries about intensionality or geometric intuitions by using a less problematic theory which can serve the mathematical ends of the original one.

Something like this can be said about the suggestions given at the end of Section 3 for a formal pattern theory—although I do not think of my view as a mathematical theory. Those who want to prove theorems about 'patterns' and find my informal exposition too imprecise might prefer to model or 'reduce' patterns by using pure sets. But the switch would be like other mathematical reductions; there would be no fact of the matter as to whether the old and new theories had the same ontology.

[10] See Giaquinto (1994) concerning such limitations.

6. REFERENCE TO POSITIONS IN PATTERNS

We can describe patterns in a number of ways. We might state how many positions a pattern has and how they are related. Or we might point to an arrangement and add that it instantiates the pattern we have in mind. Or we might introduce labels for positions in our pattern and use them to state how its positions are related to each other. But no mathematical description of a pattern—not even one by means of a categorical set of axioms—will differentiate its occurrences within other patterns from each other or from its occurrences in isolation; unless the description also states that the pattern occurs within a certain containing pattern. For mathematics only describes structures up to isomorphism, except when it describes them as embedded in other structures. This is another facet of the thesis that there is no fact of the matter as to whether two positions are the same unless they belong to the same pattern; since if a description of a pattern could distinguish its positions from those of its occurrences in other patterns then it could also distinguish the occurrences themselves.

But if there is no fact of the matter as to whether the positions in a pattern are the same or distinct from those in one of its occurrences, then there is none as to whether general or singular terms refer to the positions in the one rather than the positions in the other. This point applies even when reference is immanent and disquotational. For the predicate 'number' ('set') refers to a thing if and only if it is a number (respectively, a set). Hence 'number' refers to sets just in case numbers are sets. So if there is no fact to the matter as to whether numbers are sets, then there is none as to whether 'number' refers to sets.

Despite this, 'number' refers to numbers; and there are plenty of facts concerning what it does or does not denote. For instance, it does not refer to the even numbers only, because not all numbers are even. I am taking reference immanently, of course, which presupposes taking my language at face value. But it is difficult to understand how I could seriously wonder whether the term 'number' refers to something other than numbers. For even to begin would require rejecting the principle

> 'number' refers to numbers,

and to adopt a theory of reference in which this disquotational principle can be blocked.

Such a theory would be a transcendent one that characterized reference in terms of some relation, $R(t,o)$, applying to terms t in arbitrary languages and objects o. This would allow us to conclude such things as:

> Contrary to what Priestly thought, his term 'dephlogistonated air' referred to oxygen and not to air from which phlogiston had been removed.

A supposed virtue of such a theory is that it recognizes facts concerning the references of a foreigner's terms. But this is a dubious virtue when it comes to mathematical terms. For unless the theory is restricted it will generate facts where there are none to be had. It will do so by equating or distinguishing the positions to which foreigners refer with some of those to which we refer.

Still one might wonder how reference is possible even in the disquotational, immanent sense I use. It is simply by taking our language at face value, and not raising wholesale questions concerning its references. In short, we accept disquotational biconditionals such as:

> 'Set' refers to something just in case it is a set,

and forswear asking the further question of whether the term 'set' on the right-hand side actually refers to, say, functions or ordinal numbers. In a sense, this makes reference relative, as Quine would say, because it is relative to taking our language at face value. Within it, however, reference is absolute: I still use a two-place predicate 't refers to o'.[11]

By the same token, ascribing an ontology to our theories makes sense only relative to taking their language at face value. Thus taking the language of number theory at face value, we can conclude that its variables range over numbers, but there is no fact of the matter as to whether they range over sets. It may seem I have done away with mathematical reality or with the view that mathematical truth is a matter of the way in the world is. But I am not renouncing

[11] However, I do use different predicates for different kinds of terms. Cf. my (Sing) and (Pred1) of Chapter 9.

mathematical objects (positions) or denying that the way they are related determines what is true in mathematics. My claim is that there is enough slippage between our theories of patterns and the patterns themselves to affect reference. But it does not affect truth. For the truths of a theory of a pattern are invariant under all reinterpretations in patterns congruent in it.

7. CONCLUDING REMARKS ON REFERENCE AND REDUCTION

I want to emphasize that I do not mean to preclude formal semantics. For this is something we do within a metatheory, whose language we take at face value and use to assign references to the terms of an object language. Thus by taking set theory as our metatheory and using its terms to fix an omega sequence (for example, the von Neumann ordinals) we can interpret number theory within set theory. This does give rise to a relative sense of 'reference', since relative to this interpretation the term 'number' denotes sets, and it also gives rise to a relative sense of truth—truth under an interpretation. Reference and truth both taken as relative to an interpretation are transcendent notions, which I find clear and acceptable, so long as one forswears claims about the correctness of isomorphic interpretations.

I am not ruling ontological reforms out of court, so long as they are viewed in the spirit of my discussion of mathematical reductions. The reformist tries to persuade us that we can use one theory (say, set theory) rather than a motley. On my view, we would belie no ontological facts in making the switch. However, whether we ought to do so is another matter, which depends not upon ontology but rather upon our aims and purposes.

My own aims have not been reductive or foundational. When Frege—and Dedekind too—asked what numbers were, mathematics had neither an adequate characterization of the natural number sequence nor an axiomatic basis for number theory. Their question was in fact a demand that mathematics develop an adequate conception of the numbers. One way to do this is to clarify the foundations of number theory itself, and that clarification was a direct result of their work. Another way is to explicate numbers by means of entities for which we already have an adequate conception. Frege and his successors tried this using sets, but enough doubt has been cast upon our conception of sets for us to regard this reduction of numbers today as often not conceptually clarifying in itself.

I have been taking yet another approach to the question of what mathematical objects are. The problem is no longer one of clarifying our mathematical conceptions. Rather it is a problem of finding a philosophical interpretation of them. My suggestion that mathematical objects are positions in patterns is not intended as an ontological reduction. My intention was instead to offer another way of viewing numbers and number theory which would put the phenomena of multiple reductions and ontological and referential relativity in a clearer light. My hope is that when they are seen to arise in a fairly obvious way with respect to patterns, then they will seem more comprehensible with regard to mathematical structures.

11

Patterns and Mathematical Knowledge

1. INTRODUCTION

In Chapter 9, I maintained that we can learn about certain mathematical objects by postulating that they share the same (or approximately the same) structure as the diagrams and symbol systems we use to represent them. Physicists commonly connect newly posited particles to previously recognized phenomena through hypothesizing that the new particles have charges, masses, spins, or other physical properties. Doing so is essential to using the new particles in explanations and predictions. Positing that certain systems of mathematical objects are structurally similar to systems of physical objects is supposed to fit this methodological mould as well.

We expect newly posited physical particles to be like familiar physical objects. So postulating that they have mass, for instance, makes sense, although quite a bit of theoretical work is usually needed to link the physical properties of newly posited particles with observable phenomena. By contrast, the structural connections between systems of physical symbols and systems of mathematical objects may have seemed both mysterious and *ad hoc* to readers of Chapter 9. Why should mathematical objects have structural properties at all? And why should they have nothing but structural properties? Of course, the answers are obvious once we recognize mathematical objects as positions in patterns. What is more, if the point of positing mathematical objects is to describe certain patterns, then it is plausible to allow that systems of physical objects instantiating these patterns can inform us of properties of mathematical objects.

Developing the analogy between pattern cognition and mathematical knowledge will also make greater sense of the speculative history of the introduction of mathematical objects I presented in Chapter 9. There I speculated upon factors that could lead one to posit ideal entities existing outside space and time, but I did not explain why one would be led to view these objects as featureless positions in patterns. In this chapter I will use the idea that talk of mathematical objects is a way of talking about patterns and their positions as a basis for re-telling my story about the early development of mathematics. This story will explain why we treat mathematical objects as incomplete. It will also enable us to see how systems of numeration and computation could grow from focusing on cardinalities, why the ancients posited geometrical objects as the idealized shapes of certain physical objects but did not take numbers to be the shapes of inscribed numerals, and how the connection between proof and mathematical truth was forged. This account also suggests that mathematics can continue to develop by extending, filling in, and enriching previously recognized mathematical patterns.

Before turning to these topics I want to anticipate a possible misunderstanding of my epistemology by emphasizing two points about pattern cognition. The first is that knowledge of a pattern is quite different from pattern recognition. One recognizes a physically instantiable pattern when one can distinguish its instances from non-instances. Even chickens can learn to recognize various simple patterns, but they certainly cannot describe such patterns.

The second point is that we do not literally see or intuit patterns. After all, there is nothing to see but their positions. Seeing a pattern is more a matter of *seeing that* certain of its instances fit it or satisfy its defining conditions. To abstract a pattern from instances is neither to intuit nor to see it; rather it is a process by which we arrive at a description of the pattern by alternatively positing related positions and checking their fit against putative instances. Consider a detective who sees several strangulations over a month and thinks, 'There is a pattern to this.' Instead of waiting to see or intuit the pattern through its instances, our detective will begin to use positional or role talk, perhaps saying, 'The victim is strangled with a scarf, the murderer approaches the victim from behind, the crime is committed on a Wednesday.' In this way, in describing the pattern the detective simultaneously develops a theory of it and of the data fitting it. Moreover, even a partial description of a pattern allows one to deduce some of its other features, which can in turn be checked against purported instances. Thus the detective might be able to eliminate a suspect on the grounds of not having the size and strength the murderer needed to strangle the victim with a scarf.

2. FROM TEMPLATES TO PATTERNS

Let us now reconsider the story of the introduction of mathematical objects, and this time assume that, whether they were aware of it or not, ancient mathematicians were attempting to describe patterns. Mathematical objects eventually emerge as positions in these patterns.

We can expect that human knowledge of patterns began, like our knowledge of everything else, with experience. Experience teaches us that certain shapes and arrangements work better in certain situations than others. Thus it becomes important to recognize how things are shaped or arranged, and to recognize that this thing is shaped or configured similarly to something whose shape or configuration previously proved significant to us.

Because of their practical importance, we find ourselves driven to invent a vocabulary to refer to things exhibiting important patterns. (I am speaking now of predicates applying to concrete things, not singular terms naming universals.) Certainly the ancients had similar terms. When they needed to instruct a foreigner or novice and nothing of the right shape was readily at hand, they probably used a drawing. If so, they would have taken an important step; because they would no longer have been restricted to indicating, recognizing, or labelling present things of the same pattern and could instead represent how absent things are arranged or shaped.

I want to introduce some terminology at this point lest we confuse patterns in the concrete sense—in the sense in which we have a drawer full of dress patterns at home—with abstract patterns; dress patterns, for example, that, perhaps, nobody has either sewn or described or drawn. A paper and ink dress pattern is an instance of an abstract pattern, for it is a token, a concrete inscription, of a symbol type. However, it is also a concrete representation of a type of dress without being an instance (token) of that type—without being a dress. Right now my concern is with the use of concrete inscriptions to represent other concrete things. Reserving the term 'pattern' for abstract patterns, I will use the term *template* to refer to our usual concrete devices for representing how things are shaped, structured, or designed. Concrete drawings, models, blueprints, and musical scores are my paradigm everyday templates. Under the appropriate conventions, templates represent other concrete things, such as buildings, artefacts, or performances, which fit them in the appropriate ways. Templates are thus templates for things of the appropriate kind: blueprints are designs for buildings rather than for sculptures or performances of ballets.

One can (and I presume the ancients did) go quite some distance in our practical talk about how things are shaped, arranged, or designed without appealing to abstract patterns. Concrete templates will do the job perfectly. Using them one can learn how to design things before manufacturing them, and how to use modified designs to make discoveries about modifications in the things themselves. Without introducing abstract entities, one can be in a position to talk about possibilities, about how things might be arranged or designed or shaped. We can expect the ancients to have learned to do these things as well.

Templates have two dimensions. Syntactically they are configurations constructed according to certain conventions. Semantically they represent other concrete things by means of implicit and explicit rules of representation. So far we have only considered templates that successfully fill their representational role. But we can also use the medium (for constructing templates of a given kind) to construct configurations without any representational role, such as random doodlings on blueprint paper. Surely, the ancients did this too.

We have thus advanced the ancients from the barest recognition of the practical importance of certain shapes and arrangements to representational systems for designing previously unseen *concreta*, and thence to playful and creative attempts to explore possibilities. Before we move on, let us remember that we use language to construct templates too; for we can often describe an arrangement, shape, or design in words more accurately than we can in a drawing. Some linguistic templates might be sets of directions—instructions on how to build a serviceable lean-to, for instance. Others describe rather than instruct—a biologist's description of bee dances is an example. In mathematics, linguistic templates are our chief and most reliable methods for representing patterns.

I have not yet touched the crucial question of how experience with templates could lead us (or the ancients) to knowledge of abstract patterns. The discussion of templates is not in vain, though, because it indicates how we might have begun our initial explorations of patterns and our initial probing of the possible. It also tells us something about the local epistemology of patterns. For although initially templates represent only patterned, concrete things, they eventually also come to represent the abstract patterns that concrete things might fit. Looking at the example of a dress, we see that ultimately there will be four entities involved, two concrete ones—the dress and its template, and two abstract ones—the dress pattern and the symbol type of the template. These are related according to the following diagram.

Symbol type of the dress template	Abstract dress pattern
Type to token	Pattern to instance
Template	Dress

People who countenance abstract patterns might also associate abstract patterns with templates along the lines depicted by the diagram given above. If so, they could construct and study templates to gain information about patterns. Finally, having bonded patterns and templates in this way, it would become plain to them that to show that a specific abstract pattern for, say, houses exists, it suffices to exhibit a blueprint for houses of that pattern. We can thus see how the ancients might have come to believe that paper and pencil constructions and computations could provide information about the mathematical realm.

Another important point to notice is that the sort of sophisticated use of templates we have been contemplating depends upon developing complex syntactic systems (such as place notations for the numbers and geometric diagrams), conventions and rules for manipulating these systems, semantics by which they can represent concrete entities, and at least some sense as to whether a given syntactic configuration counts as a coherent representation. I had norms like these in mind in Chapter 9 when I remarked that the use of symbol systems might have led to the emergence of deductive reasoning and proof as a method of mathematical discovery and justification. Of course, all this would contribute to easing the step towards full-blown mathematical theories with commitments to mathematical entities. Or, as I would now say, it would lead towards theories of patterns committed to positions.

We can now see why, when positing mathematical objects, it would make sense also to posit structural analogies (or isomorphisms) between the elements of diagrams and computations and configurations of positions in patterns. For the point of positing positions is to describe just the structure itself, unencumbered by the extraneous features of a diagram or concrete exemplar. By its very nature a pattern should be structurally analogous to its instances and its template. In positing that it is, one simply projects onto structures features already attributed to templates.

You might agree that this makes sense for geometric patterns and diagrammatic templates, and still wonder about the arithmetic case. What structural similarities are being posited here? Why, for instance, don't we project the shapes of numerals onto numbers? The answer to the last question is obvious, of course, since the structures numerals represent are cardinal and ordinal types instead of shapes. But this very answer makes the first question all the more pressing, since no transparent connection between individual numerals and these structures explains why one might plausibly assume that operating with the former can yield information about the latter. We can figure out a plausible connection, however, by looking at how numbering systems and computational principles might have developed.

At the most primitive level people use bags of pebbles or tally sticks to count and record counts. In seeking efficiency they will quickly come to use special pebbles or marks for key tally points, just as we mark through four slashes to signal each fifth item on a tally sheet. Further refinements of this system might lead them to aggregate marks in rows containing, say, at most 10 fifth tally symbols, as in the following row:

~~||||~~ ~~||||~~ ~~||||~~ ~~||||~~ ~~||||~~ ~~||||~~ ~~||||~~ ~~||||~~ ~~||||~~ ~~||||~~.

Then each filled row would indicate a total of 50 items counted. The Roman numeral system seems to be an outgrowth of an approach like this. On the other hand, we need not use special symbols for counts of 5, 10 , etc. if we simply arrange our tally marks in patterns. For instance, instead of writing

• •• ••• •••• ••••• ••••••

when counting from 1 to 6, we might write

```
     •   •   ••   ••   •••
•    •   ••  ••   •••  •••
```

which is not only more compact but also suggests how the numbers are related to one another, as we shall see in the next section. Another good feature of this change from the first linear dot notation is that it preserves the cardinality of each group of dots and, thus, can represent any information recorded by the former in a more perspicuous form. Clearly coming to think of numbers as something like cardinalities depends upon recognizing that such rearrangements preserve the information the marks were designed to record.

Because our current decimal notation uses different simple symbols for 0 to 9, it easier to read and write than either the dot or Roman notations. It also works immeasurably better in calculating, because it is a *place notation*, that is, one in which both the digit and the place it occupies within a compound numeral carry information. But while our notation is better suited for store-keepers, it obscures the connection between computations with it and facts about the numbers. It also obscures the fact that in the more primitive systems of numerals, such as the dot system above, the individual numerals serve as templates for cardinalities. The dot numeral for the number n has exactly that many distinguishable parts.

So let us return to the more primitive level, and see how one might calculate with a system of *linear* dot templates for representing cardinalities. To 'add' two templates we simply juxtapose them. To 'subtract' a smaller template from a larger one we line up the smaller one below one end of the larger and then cross off every pair of vertically aligned dots. Finally, to multiply two dot templates we lay one horizontally and the other vertically so that their endpoints overlap. Then we fill in the space below the horizontal dots until we obtain a rectangular array, which we then rearrange in the standard linear form to yield the 'product' of the original templates (see Figure 1).

Using the dot system I have just described one can quickly come to appreciate that addition and multiplication are associative and commutative, that multiplication distributes over addition, and that 1 is a multiplicative identity. Indeed, this dot system provides a much

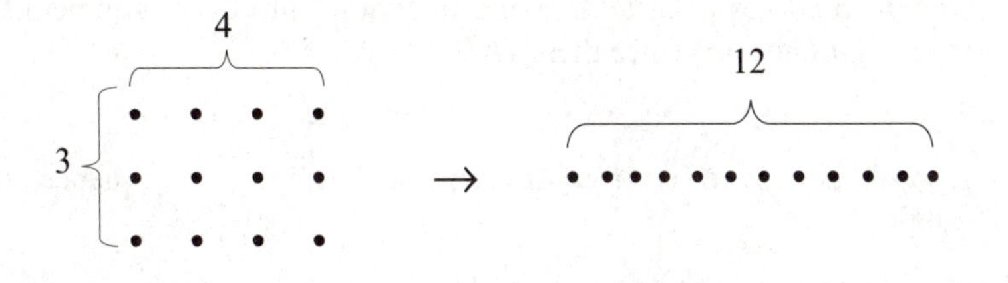

FIGURE 1

better understanding of why those principles hold than any explanation we are usually given.

To convince ourselves that these methods are sound, we might try them on a number of examples. But it would be much more effective for us to observe that dot addition literally aggregates disjoint collections of dots, that dot subtraction deletes an appropriately sized sub-collection of dots from a collection of dots, and that dot multiplication replicates a collection the appropriate number of times and then forms the aggregate collection of dots.

Because individual dots have no features for distinguishing them from other dots, each dot numeral, that is, each linear collection of dots, represents all collections of things of its cardinality. Thus it also represents itself *qua* collection of dots. Furthermore, dot arithmetic is a method for operating on dot numerals that produces dot numerals having as many dots as one would get by physically 'adding', 'subtracting', or 'multiplying' the corresponding collections of dot tokens. Consequently, dot arithmetic is also a method for operating on templates which produces templates for collections that have the cardinalities one would obtain from 'adding', 'subtracting', or 'multiplying' the collections themselves.

If the ancients introduced abstract patterns corresponding to dot templates, then it would seem that each numeral (dot template) would correspond to a cardinality *qua* pattern, and that the individual dots in the numeral would correspond to the (unrelated) positions of a cardinality. Numbers would emerge as cardinality patterns instead of positions in the number series. But it is not hard to see how the ancients might have moved from dot numerals for cardinalities to the numbers series. For each dot numeral is obtained from its predecessor by adding one more dot. Thus the dot numerals form an infinite sequence themselves:

•, ••, •••, ••••, •••••, ••••••, etc.

It would be easy to slip from this sequence to an infinite sequence of dots:

•••••••••••••••••••••••••••, etc.

in which the *n*th pattern of dots of the first sequence is represented by a position, the *n*th dot, of the second sequence. The whole sequence would be the pattern relating the individual numbers, initially treated as patterns, and now treated as positions.[1]

We can now see why it would have been plausible for the ancients to assume that computational principles originally designed for dot arithmetic could furnish information about numbers, now thought of as positions in the number sequence. For there is an obvious way to convert the principles of dot arithmetic into rules for adding, subtracting, and multiplying numbers. For example, to add two numbers, one simply dot-adds the dot patterns the numbers now represent and assigns the number representing the resulting dot pattern as the sum of the two numbers.

3. FROM PROOFS TO TRUTH

One of the major problems facing mathematical realists is to explain how mathematical methods, particularly those based upon computing and proving, could generate information about the mathematical realm. In Chapter 9 I said that mathematicians learn about this realm through appealing to structural similarities between various abstract mathematical structures and physical computations and diagrams. Lately I have been using structuralist insights to give a deeper account of these similarities. But I am not claiming that we can mirror the entire mathematical universe with diagrammatic templates. After all, the latter are always finite while the former is vastly infinite. Rather, just as natural science tests its theories against observations, mathematicians can test their theories by carrying out

[1] If we think of zero as the first number, then the *n*th member of the original sequence would be represented in second sequence by its $n + 1$st.

certain physical computations and constructing certain templates. I am not the first, of course, to suggest that mathematicians can and do obtain evidence for higher-level theories through results belonging to more elementary levels.[2] What I have been adding is a connection between certain elementary mathematical results and physical operations that we can perform.

It is well known that axiomatic geometry arose as an attempt to systematize a body of results concerning measurements and diagrams. Of course, I would urge that we interpret these as obtained by reflecting on templates for geometric patterns. It is likely that the Greeks regarded many of these results as incontestably true and not needing further proof, and, consequently, felt free to use them to test the adequacy of a proposed set of axioms.[3] Now one might wonder how a body of results sufficient to produce number theory could originate from reflecting on numerical templates, since practical computations are not likely to suggest facts about the sums of finite sequences of numbers, the infinity of primes, or the prime factorization theorem, which were at the heart of Greek number theory.

Greek number theory probably began in the contemplation of dot templates. For the Greeks used shape terms to classify numbers. They counted some numbers as triangular, others as square, and others as oblong. To see how they might have reasoned, suppose that we use dot templates to represent 1 to 4 in Figure 2.

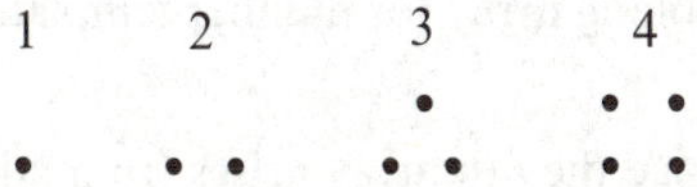

1 2 3 4

FIGURE 2

Let us understand 1 as both a degenerate square and triangle. We can see that the dot array representing 2 is oblong, while that representing 3 is triangular, and the one representing 4 is square. Just

[2] Gödel, Quine, Kitcher, and Maddy have all held that postulates of higher branches of mathematics can be justified in terms of their lower-level consequences. See Gödel (1963), Kitcher (1983), Maddy (1990), Quine (1986), 400.

[3] I expect that their ability to visualize constructions with templates had a role in making some general statements obvious. For further work on how visualizing can lead us to see certain mathematical claims as obvious, see Giaquinto (1992).

these simple diagrams can prompt more general questions. Which numbers are triangular? Which are square? Which are oblong? To find this out, let us write down sequences of the variously shaped dot representations and see whether some patterns emerge. Starting with the oblong numbers we have the sequence shown in Figure 3.

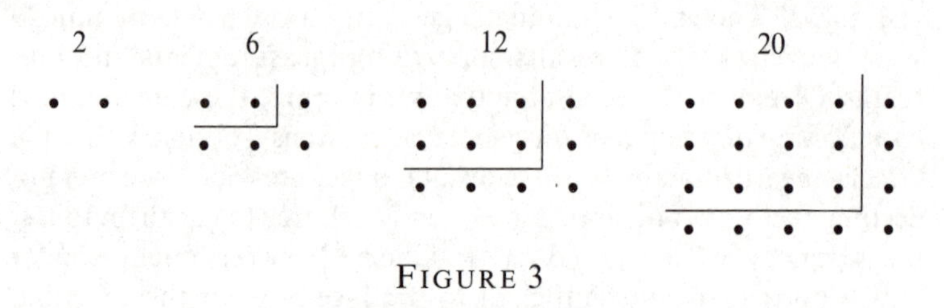

FIGURE 3

Notice that each oblong array is generated from the previous one by adding a new column of dots to its right and a new row of dots to its bottom. These examples start with an array one dot high and two dots wide, move to an array two dots high and three dots wide, and then on to one three dots high and four dots wide. The general pattern should be plain: the nth array is n dots high and $n + 1$ dots wide and is obtained from the previous one by adding n dots arranged vertically on the right and then n dots along the bottom. Thus we have the following result:

(1) The nth oblong array (or number template) contains $n(n + 1)$ dots.

Furthermore, since the nth array arises from adding $2n$ dots to the previous array and the series starts with a two-dot array, the nth array contains as many dots as the sum of the first n even numbers. Thus we also obtain:

(2) The sum of the first n even numbers equals $n(n + 1)$.

If we express this in modern notation as:

(3) $2 + 4 + 6 + \ldots + 2n = n(n + 1)$,

we can easily prove it by mathematical induction.

It is interesting how easily our attention has shifted. What began as a quest for an arithmetical characterization of the oblong numbers ended with a formula for the sum of the first n even numbers.

This suggests that we now ask two sorts of questions: what is the formula for numbers having a dot template a given shape? What sums do they represent? This will quickly lead us to use the dot technique to find that the square numbers are characterized by the formula n^2, which is also the sum of the first n odd numbers, i.e.

(4) $1 + 3 + 5 + \ldots + 2n-1 = n^2$.

What about the first n numbers? Well, each oblong array contains as many dots as the sum of the first n even numbers; so if we divide these arrays in half by drawing diagonals through them and erasing all the dots below the diagonal, then each new triangular array will have half as many dots and can be obtained from the previous one by adding half the dots to it as were necessary to obtain the oblong from which it was produced. This means that the nth triangular array will contain $\frac{1}{2}(n(n + 1))$ dots and can be generated from the previous array by adding n dots to it. This means that

(5) $1 + 2 + 3 + \ldots + n = \frac{1}{2}(n(n + 1))$.

We could easily prove this by mathematical induction. But we can derive it more quickly from (3), since (3) may be rewritten as

(6) $2 \cdot 1 + 2 \cdot 2 + 2 \cdot 3 + \ldots + 2n = n(n + 1)$,

from which (5) follows by dividing both of its sides by 2. Furthermore, the last method for obtaining (5) can be generalized to yield a formula for the sum of the first n multiples of any number p, namely,

(7) $1p + 2p + 3p + \ldots + np = \frac{1}{2}(pn(n + 1))$.

Our modern arithmetical notation gave us an advantage over the Greeks and allowed us quickly to find (7). But the Greeks could have arrived at the same result by reasoning about dots. Suppose that each dot in the dot triangles used in the derivation of (5) is replaced by p dots. Since we have increased the dots in each array p-fold, the nth array of dots will represent the sum of the first n multiples of p. Since it contains $p[\frac{1}{2}(n(n + 1))]$ dots, this number also equals the sum of the first n multiples of p.[4]

Thus we can see how working with templates could generate a body

[4] See Heath (1981) for further discussion of the dot proofs.

of results which could eventually evolve into a systematic theory of numbers. Of course, finite templates can represent only initial segments of an infinite number sequence. Thus we must turn to linguistic templates, that is, go axiomatic. Let us suppose, then, that we want to develop a pattern which is, so to speak, an infinite dot sequence and the union of all its initial segments. Then not only will the usual axioms for number theory prove their worth in generating results such as those obtained above, they will also strike us as obviously true of the pattern we have in mind. Thus we can see how starting with initial dot templates for finite cardinalities could lead us to introduce both an infinite number sequence and theorems about it.

I began this section by raising the general question of how proofs and computations generate information about the mathematical realm. The number-theoretic example gives us a foothold on an answer, but we cannot simply generalize from it. Today we calculate using arabic numerals and the rules we learned in school. If we seek a more basic explanation of why they work, we can appeal to theorems of some axiomatic number theory, or alternatively we can explain our current rules in terms of dot arithmetic. For we can easily recover school arithmetic from either of these bases and appropriate definitions. Either way, we can reveal information-bearing connections between our computations and the natural number sequence. For in proving theorems in axiomatic number theory we reason from obvious features of the natural number sequence, and in doing dot arithmetic we manipulate templates isomorphic to its initial segments.

Unfortunately, we cannot plausibly underwrite most computations by appealing to manipulations with templates. The rules for computing the values of derivatives, trigonometric expressions, or transfinite polynomials, for example, did not arise from manipulating diagrammatic templates for derivatives, trigonometric functions, or transfinite ordinal numbers, since there are no such templates. Instead these rules are founded upon theorems of the pertinent theories. So here there is no relatively straightforward connection between the computations and the patterns they are supposed to concern.

In such cases there are connections between proving a theorem and the pattern (or patterns) the theorem concerns. Most proofs do not take place within an explicitly formulated axiom system, but those that win acceptance must be based upon uncontroversial premisses. We forge a connection between a proof and a pattern when the proof's premisses state uncontroversial features of the pattern in question. This happens in the number-theoretic justifications of school algorithms, and it happens, for example, when we prove a theorem about the real numbers using only premisses that state uncontroversial features of the real number sequence. For here the information-bearing connection between proving a theorem concerning a pattern and the pattern itself is just a special case of the connection we have to things generally when we reason about them.

Now two problems with this approach come to mind. The first is that some proofs of theorems about some relatively simple structure proceed by embedding the structure in some more complex structure. Proofs of theorems in elementary number theory that appeal to premisses from real or complex analysis do this; so do consistency proofs for number theory based upon epsilon-zero transfinite induction. Although this interesting phenomenon is worth further examination, I don't think we need worry about it here. When we prove a theorem of number theory using, say, complex analysis, we appeal to information about a larger pattern in which the natural number sequence occurs. This is no different from giving a number-theoretic proof of a theorem that is restricted to some initial segment of the number sequence. We do this, for example, when we use the fact that 1 is the only number that divides every number and its successor in order to prove that it is the only number between 0 and 100 that divides each of these numbers.

The second problem is not so easily set aside. This is the problem of explaining how we know that the premisses stating supposedly uncontroversial features of a pattern to which the proof appeals are true of the pattern in question. The short answer is that we implicitly or explicitly posit that they are true of it, just as we posit that the dot templates mirror features of (initial segments of) the natural number sequence.

But I must elaborate a bit on this answer, since one might remark that I am saying that, in effect, the premisses to which we appeal in proving the theorem 'implicitly define' the pattern or class of patterns to which the theorem pertains. Now I have no problem *per se* with calling such premisses (or a more condensed set of axioms from which they might be derived) an implicit definition, so long as this is not taken to imply that the premisses are known a priori in some absolute sense. Of course, the axioms constituting the clauses of an implicit definition are trivial consequences of this definition. Thus it is a matter of definition that they characterize the pattern they help specify. But we are also supposed to know that there are positions forming such a pattern, and this is not a matter of definition. The epistemological discussion I have been developing in this chapter takes place within the context of my holism cum a local conception of mathematical evidence. In postulating positions forming a pattern with certain basic features we enlarge our local conception of mathematical evidence. Now it may happen that we appeal to purely mathematical considerations in justifying introducing these postulates. Furthermore, once the postulates are in place they not only serve as evidence for theorems but also are candidates for Euclidean rescues. Thus from the local point of view they may appear a priori. However, this point of view is founded on a global holism according to which no mathematics is known a priori.

Consider in this light my suggestion that the natural number sequence might have been introduced as the union of all its initial segments. Since it is an infinite pattern, mathematicians must have used linguistic templates to describe its basic features. Let us imagine them trying out various descriptions of the pattern they were after. These attempts might have misfired through proving inconsistent, or through failing to reproduce appropriate features of the dot patterns with which they began. (For example, they might have failed to provide for a commutative form of addition.) In either case, they would have revised their descriptions. Let us suppose that eventually they are able to reproduce the features they desire without encountering any inconsistency. Still, they could have no proof that their 'axioms' are consistent, since any premisses to which they might appeal would be as questionable as the consistency of their axioms.[5] The decision to allow them to serve as a part of the evidential basis for further mathematical developments would have to be made from the more global perspective of the benefits to mathematics and science generally.

We began with asking how one might get information about a

[5] Against those who might use Gödel's theorem at this point, Detlefsen has argued that it may be possible to give a consistency proof for number theory using less questionable premisses. But my point is that even weak systems, such as primitive recursive arithmetic, are committed to 'potential' infinities, whose existence cannot be proved without already assuming some (potential) infinity. See Detlefsen (1986).

pattern through proving theorems. I answered that one would reason from premisses true of the pattern. This prompted the further question of how one might know the premisses are true of the pattern, to which I responded that they constitute an 'implicit definition' of the pattern. Now this might explain how one might know that one has specified *some* pattern, but it does not explain how one knows that one has specified the pattern one seeks. I think the question of whether one has got the pattern one sought makes sense only in so far as prior to trying to specify the pattern one has a reasonable characterization of its features. If one has already uniquely characterized it, then in implicitly defining it one simply exchanges one description for another. This could happen if one described a pattern as a sub-pattern of some pattern and asked for a characterization of it in isolation from its larger context. For example, one might ask for an abstract characterization of the structure of the integers modulo 13. But usually we begin with partial and vague descriptions of the pattern we seek. Then it might turn out that several incompatible pattern specifications can meet our preconditions. In this case, I see no reason why there should be a fact of the matter as to which of the alternative patterns is the one we seek. For it is difficult to see what could count as evidence for deciding the matter. This is reinforced by the consideration that even adding to our preconditions that the pattern must be the simplest, most natural, mathematically fruitful, and so on, may not settle the matter.

Allowing that we might implicitly define patterns raises the worry that this will make the theory of the pattern analytically true. For the theorems follow logically from conditions specifying their subject-matter. Thus, for example, one might think that 'The natural ordering of an omega sequence is discrete' is analytic, because it is a logical consequence of an explicit definition of 'omega sequence'. One can, of course, convert the conditions positing a structure (implicitly defining it) into an explicit definition of that type of structure.[6] Furthermore, replacing these conditions by non-equivalent ones suffices for changing the subject, so to speak. However, the doctrine of analytic truth claimed that some truths are true in virtue of the way one's language is while others are true in virtue of the way the world is. On the structuralist view, the theorems of a branch

[6] This observation originates with Frege. See the discussion of the Frege–Hilbert controversy in Resnik (1980).

of mathematics are supposed to be true of the structure (or structures) they describe. *Part* of that claim, I'll grant, can be paraphrased as the claim that they follow from the clauses defining the structure(s) in question. But the other part of the claim is that there are structures (or at least appropriately related positions) of the kind in question. This existential claim is not a logical consequence of the definition of the structure in question. (That an omega sequence has a first element follows from the definition of omega sequences, but that there are things forming an omega sequence does not.) Thus combining structuralism with the doctrine of implicit definitions does not automatically make mathematics analytic.[7]

4. FROM OLD PATTERNS TO NEW PATTERNS

We saw earlier how the study of the natural number sequence might have arisen through the desire to collect separate numerical patterns together into one encompassing pattern. Euclidean geometry could have developed in a similar way. It might have begun as the study of certain geometric designs and shapes considered in isolation and evolved into a more comprehensive and systematic framework for treating all these shapes and designs within the context of one space. Collecting patterns originally treated in isolation into one pattern is one way in which mathematics arrives at new patterns. In the process, previous patterns may be represented as positions in the new pattern. We have seen how the individual cardinalities can be represented by positions in the natural number sequence, and, in the last chapter, how the number sequence itself can be represented by a position in a set-theoretic hierarchy.

Mathematics also arrives at new patterns by 'extending' and 'filling in' the patterns it already recognizes. Histories of mathematics frequently use these metaphors in describing developments of the number system. Thinking in terms of patterns gives the metaphors a new twist. For example, it suggests that one might introduce the integers by taking two copies of the natural number sequence,

[7] Of course, it might be open to other versions of structuralism to hold that the claim that things exist having an implicitly defined structure is also analytic. Interestingly, although Hellman asserts that it is logically possible that various implicitly defined structures are instantiated, he does not take his assertions to be analytic: Hellman (1989), 32.

reversing one, and tacking it on to the front of the other, or that one might arrive at the rationals by filling in the gaps between the positions forming the integers. We can also see why set theorists would tend to think of the iterative hierarchy as open-ended, and be attracted to axioms positing previously unrecognized sets rather than ones closing off the hierarchy. For it is plausible that any pattern can be extended by just adding new positions to it.

However, combining, extending, and filling in do not account for all the new patterns that occur to mathematicians or even for the concepts they use to describe features of the patterns they already recognize. Staring at dot patterns has a limited role in promoting number theory. It could have led the Greeks to recognize the prime numbers; since the odd primes are neither square nor oblong, yet every composite number is. But it is unlikely that it led them to conceive of primes as the ultimate factors of every number or to dream up Euclid's proof that there are infinitely many primes. For the distribution of primes in the natural number sequence is irregular, and the individual primes themselves conform to no geometrical pattern that can be transformed from one prime to yield the next. Two is oblong, three is triangular, and five is trapezoidal (as is every greater odd number).

Instead, the idea that primes are ultimate factors must have emerged from patterns the Greeks observed while calculating. For once one begins to list factors of numbers, the prime number concept is inevitable. So, too, is the conjecture that every number is prime or may be completely factored into primes. For this follows easily from the definition of prime number and the result that every composite number is a product of finitely many smaller numbers.[8]

We see, then, that mathematicians might look for patterns even when operating at one or more steps removed from the patterns which form the current subject-matter of a mathematical theory. The Greeks factored some numbers into primes. They saw a pattern emerge: factoring leads to smaller and smaller factors until we reach primes. This probably led them to conjecture that the same holds for all numbers. Then by returning to their examples, they might have extracted general properties upon which to base a proof.

[8] It is much harder to prove that the prime factors of a number are unique up to their ordering in the product, although a few examples would quickly convince one that it is true.

Practical and scientific experience has often suggested new patterns and generated new branches of mathematics. But it is important to note that mathematics also produces new patterns and theories by reflecting on its own activities. Patterns arise in proofs, calculations, and solutions to problems. Describing them in terms of their positions leads to new theories of new mathematical objects, such as the theories of equations, proof theory, the theory of computability. The process can go on and on.

12

What is Structuralism? And Other Questions

1. INTRODUCTION

In this chapter I explore several issues raised in the last two chapters, and attempt to unify structuralism with the book's other theses. I begin by clarifying the 'fact of the matter' locution, which figured so prominently in Chapter 10. With this in hand I turn to explaining why I maintain there that there is no fact of the matter as to whether the patterns various mathematical theories describe are themselves mathematical objects (positions in patterns). Next I discuss the relativity of key structuralist concepts, for example, sub-pattern and pattern equivalence, to the choice of a background notion of definability, and consequently to the choice of a background logic. I then explore the possibilities for formulating structuralist versions of mathematical theories, and ask whether structuralism is a foundation for mathematics or a philosophical account of it.

I then turn from questions about my structuralism *per se* to the task of squaring it with my realism, my disquotational approach to truth, and my rejection of a rigid abstract/concrete distinction. I conclude with a summary of the views developed in this book.

2. ON 'FACTS OF THE MATTER'

What does it meant to say that there is a fact of the matter that *so and so*? I have borrowed the locution from Quine's writings on reference and translation; but I will not attempt to explicate his use of it, and will merely try to give a more precise account of what I mean by it.

For me, the notion of a fact of the matter is immanent, that is, limited to specific languages. I recognize no language-transcendent fact-of-the-matter predicate. Thus the notion of fact of the matter really splits into a family of notions: *there-is-a-fact-in-L-of-the-matter-as-to-whether*, where 'L' is a place-holder for a specific language, and not a variable. Suppose then that we have a fixed language L and a sentence *p* of L. Then there is a fact of the matter as to whether *p* in L just in case *p* is true or false (in L). (Recall that truth, for me is also immanent.) I allow, however, that there may be a fact of the matter of some *p* even when we may never be in a position to determine whether *p* is true or false.

On this view, when I deny that there is a fact of some matter, for example, as to whether numbers are sets, or whether so-called Collins Mountain is a genuine mountain, or whether electrons have positions when not in an eigenstate for position, I am tacitly referring to certain sentences in our language, and at the very least I am denying that they have truth-values. On the disquotational approach to truth I took in Chapter 2, one can deny a truth-value to a sentence by excluding it from the field of one's truth-predicate or by weakening one's logic to the extent of withdrawing those instances of excluded middle sufficient to generate the unwanted cases of bivalence.[1] But one might take an even more radical stance and deny facts of the matter by claiming that the offending sentences should be banned from a proper language for science. This would seem an appropriate way to deny that there is a fact of the matter as to whether exposing phlogiston to sunlight dispels it from a container of air or whether the Absolute permeates the True. (It is also a plausible reading of Quine's denials of facts of matters when it comes to synonymy and meaning.)

In the previous chapters when I said that there is no fact as to whether, for example, the numbers are sets, I had no specific language in mind, but it now seems reasonable to reconstruct my discussion as tacitly referring to sentences in the language I was using, to wit, ordinary English supplemented with mathematical notation and terminology. Now one reason that I gave in Chapter 10 for thinking that there is no fact of the matter as to whether numbers are sets, is that mathematics neither affirms nor denies this claim. Indeed, it is not clear that the mathematical part of our language even contains the sentence 'numbers are sets'; for whether it does

[1] Suppose *p*, then by disquotation for truth '*p*' is true. Suppose not *p*, then similarly '*p*' is false (not true). Thus by excluded middle '*p*' is true or false.

depends upon the particular formulation of set theory one uses.[2] But this is not really to the point, since in the fuller language I was using one can freely combine its general terms to form sentences of the form 'the *X*s are *Y*s'. So we may fairly conclude that my language contained the sentence 'numbers are sets' and other identities between various mathematical objects. Hence, in saying that there is no fact of the matter as to whether certain mathematical objects (or positions) are identical to others, I was tacitly denying truth-values to this and related sentences.

I am committed to more than denying truth-values to whole sentences, such as 'each number is a set'. For I have also held that for any number (or more generally, any thing) there may be no fact as to whether it is a set, i.e. whether 'set' is true of it.[3] Unfortunately, it is not clear that I can satisfactorily follow through on this by merely restricting my truth-predicate or truth-theory. Plainly, banning the term 'set' altogether or restricting the satisfaction (true of) relation so that it does not apply to 'set' is too radical. Furthermore, restricting *satisfaction* would mean specifying restrictions on objects, and these are likely to be circular. For instance, it seems that we will require a condition such as:

> The pair (x,'set') falls within the field of *satisfies* iff there is a fact as to whether an x is a set,

which presupposes the very idea we are trying to make more precise. In light of this, I think it is best to restrict logic so that excluded middle does not apply generally.[4]

In the end, then, implementing my claim that there is no fact of the matter as to whether numbers are sets involves rejecting the universal applicability of classical logic. I don't think this is as much of a reform as it might initially appear. Our actual practice in this instance is prima facie classical. We seem prepared to affirm, for example, that 2 either is or is not $\{\{0\}\}$. We are also initially disposed to hold that Hamlet either did or did not have an ingrown toenail or that John Stuart Mill was or was not a great philosopher.

[2] It depends not only upon whether 'number' is a primitive or defined term, but also upon whether defined terms count as part of the language.

[3] One might say that I rejected both *de dicto* and *de re* facts of the matter.

[4] One way to do this would be to use intuitionist logic for this area of discourse. Note that I am not espousing mathematical intuitionism, for I am not suggesting restricting excluded middle *within* individual mathematical theories.

Despite this, our commitment to disjunctions, such as the last two, which involve fictional or vague terms, is not strong. We feel no real need to decide which of their disjuncts are true, and we can easily give up the disjunctions themselves with little consequence to our overall theory of the world.

Nothing in science or mathematics, as ordinarily understood, counts as evidence for or against numbers being sets. Obviously, given the philosophical attention directed to the issue, the matter is not entirely inconsequential, but the only theories that seem to be directly affected by a decision one way or the other are philosophical theories. Thus letting excluded middle lapse in these cases seems to have the benefit of resulting in a simpler philosophical account of mathematics without generating untoward reverberations elsewhere.

The basis for deciding whether to recognize a fact of the matter varies with the supposed matter at hand. The decision whether to recognize a fact of the matter of Mill's philosophical greatness depends upon finding a convenient way of dealing with the vagueness of the term 'great philosopher'. On the other hand, that there is no fact of the matter of the position of a particle which is in a superposition of position states seems to be a deep and hard-won result of physical theory. Finally, my *dicta* concerning identities between various kinds of mathematical objects are based upon likening them to positions in patterns. Thus they derive from an insight concerning the nature of mathematical objects. They are, if you will, consequences of my philosophical premisses.

3. PATTERNS AS MATHEMATICAL OBJECTS

In Chapter 10 I noted that the typical mathematical theory excludes the structure it is supposed to describe from its universe of discourse, and consequently does not recognize it as a mathematical object. For example, most formulations of number theory have individual variables for numbers, but no *individual* variables for number-theoretic relations, functions, or the number sequence itself (and no variables for these at all if the formulation is first-order). These theories have the wherewithal for asserting that there are numbers but not for asserting that there is a number sequence.[5] This is why I

[5] At least they cannot state that the number sequence exists as an object; most

described these theories as treating the individual numbers, but not the number sequence itself, as *positions* in the structure (or structures) which they attempt to describe. It is also why I would describe them as treating the number sequence as a structure.

Though these theories cannot treat the natural number sequence as a position, other theories can. For instance, set theory can represent it as a set (*N*,*S*). This example typifies mathematics. None of the usual mathematical theories can treat the very structure(s) it attempts to describe as a position in that structure, because so doing would require its universe of discourse to be a member of itself. Yet, by positing more positions and extending the original structure, one can usually represent it as a position in a new structure. Something like this holds even for ZF set theory. It cannot countenance any of its models as a set, but some of these models can be represented as sets in extensions of ZF models.

I also noted that although we may identify or represent the natural number sequence with the set (*N*,*S*), there is no fact of the matter as to whether they are the same. One reason for maintaining this is familiar: namely, that more than one set can serve as the natural number sequence—(*S*,*N*) will work—and more than one set can serve as the successor relation. But now that I have clarified my use of 'fact of the matter', we can appreciate an additional consideration: for there to be a fact of the matter in this case, we need a language in which to formulate the claim that the sequence treated by number theory (without sets) is identical to a certain set. But number theory does not refer to the number sequence (or at least does not refer to it as an individual), while the contemplated language would refer to it as an individual in asserting its identity with a set. This would presuppose (the fact) that there is an object treated by number theory identical to what this language calls 'the number sequence', and so we would be off on a regress of presuppositions. Of course, the very language I am using contains expressions such as 'the set (*N*,*S*)', 'the natural number sequence', 'the pattern which number theory tries to describe by referring only to its positions',

second-order number theories can posit a successor relation, which some people would identify with the number sequence itself. So as not to confuse entities of different logical types, I will use 'entity' as a generic term for anything of any logical type, and follow custom in reserving the terms 'object' and 'individual' for things of the lowest logical types (first-order entities).

and so on; but for the reasons adduced in the last section, the mere presence of these terms need not commit me to recognizing a fact of the matter as to whether these terms refer to the same object. For the same reason, my bringing the numbers, the number sequence, and the positions of various structures under my own theoretical umbrella does not commit me, after all, to there being a fact of the matter as to which entities the numbers are. As above, I resist such matters of fact by appropriately restricting excluded middle.

A similar point applies to my frequent talk of *treating the numbers* here in one way and there in another way. I was referring to differences in various explications of informal but mathematical number-talk.[6] I was assuming, of course, that each of these explications would attempt to describe a structure or structures, and simply denying that there is a fact of the matter as to which of them correctly formalizes the informal talk. But let us suppose that the informal mathematical talk already describes positions in a structure, to which the phrase 'the number sequence' refers. Then what I deny is that there is a fact of the matter as to whether the positions of the various structures introduced as explications are identical with whatever our informal talk of numbers refers to.

So far I have been pointing out that some mathematical theories explicate important mathematical structures by identifying them with mathematical objects, while others do not countenance them as entities at all. Because of this, I concluded that there is no fact of the matter as to whether patterns are mathematical objects or entities of any kind. But don't I take that back in my own theorizing? For I refer to individual patterns, apply predicates and relational terms to them, and in my definitions of pattern occurrence, sub-pattern, and pattern equivalence I even existentially quantify over patterns.

Towards resolving this difficulty, let me compare the problem of explicating my informal talk of patterns with similar difficulties in explicating other informal notions. Take the familiar fact that words for numbers, properties, and relations can often function both as adjectives and as nouns. Here are some reminders:

(*a*) A yellow strip marks the centre of the road.
(*b*) Yellow is more noticeable than green.
(*c*) Yellow is a colour.

[6] In the Carnap–Quine sense of explication as elimination.

(*d*) Three cats are on the mat.
(*e*) 3 is less than 4.
(*f*) *Less than* is the converse of *greater than*.

Judging from (*a*) yellow cannot be an individual, yet this is how (*b*) and (*c*) treat it on their face-value reading. Similarly (*e*) treats 3 as an object while (*d*) does not; (*f*) treats the *less than* relation as an individual but (*e*) does not. Philosophers explicate examples of this sort by preserving the apparent logical form of some of the cases while paraphrasing away the conflicting ones. Thus, having decided that numbers are objects, Frege argued that we can paraphrase adjectival uses of number words, as in (*d*), in favour of nominal uses. Nominalists, on the other hand, would favour the other direction, and might paraphrase (*f*) by:

(*g*) Something is less than another just in case the latter is greater than the former.

Furthermore, sometimes avoiding the offending locutions involves going metalinguistic. Thus nominalists might offer:

(*h*) The term 'successor' is a two-place predicate

as a substitute for the claim that successor is a relation.

Now I am not about to develop a mathematical explication of my informal discussion of patterns. But were I to undertake this project I think it would be appropriate to begin by dividing my pronouncements into two sorts. Those of the first sort concern patterns alone and their relationships to one another, and they include my definitions of sub-pattern, pattern occurrence, and pattern equivalence. This body of doctrine is ripe for mathematical treatment through a mathematical theory of patterns. Notice, however, that a mathematical theory of patterns would treat patterns as positions in a pattern of patterns, and thus as mathematical objects.[7]

Accepting patterns themselves as mathematical objects clashes with many of the other things I have said—for example, my claim that there is no fact of the matter as to whether the pattern number theory studies is a mathematical object, and the like. But claims of this type can be regarded as shorthand for the more elaborate explanations I presented a few paragraphs back. Thus in order to avoid

[7] I will say a bit more about prospects for such a theory in Section 5 below.

speaking of the pattern treated by the number theory, I might replace the last example by the following elaboration:

(1) Number theory describes how the numbers are related to one another without stating anything more about their nature;
(2) It treats the numbers as if they were positions in a pattern;
(3) Its variables range over numbers only;
(4) There is no fact of the matter as to whether the numbers are members of the set (N,S).

(This presupposes, of course, my earlier explanations of the 'fact of the matter' and 'treats' locutions.)

Notice that (1)–(4) are meta-level comments about number theory, and only (2) speaks of patterns and positions. Were a mathematical theory of patterns at hand, we might regard (2) as asserting that number theory could be reduced to or explicated through it. As it stands, it merely asserts an analogy between the numbers and positions in a pattern.

Of course, (1)–(4) are not mathematical substitutes for the claim that there is no fact of the matter as to whether the pattern number theory treats is a mathematical object. They are merely the results of my effort towards making my philosophical discussion more explicit. For now I will leave the question of how literal and precise an acceptable philosophical theory must be, but I promise to return to it in Section 5 below.

4. STRUCTURAL RELATIVITY

In thinking about formulating a theory of structures we must take into account a phenomenon I will call *structural relativity*: the structures we can discern and describe are a function of the background devices we have available for depicting structures. This relativity arises whether we think of patterns and structures as a kind of mould, format, or stencil for producing instances, or as whatever remains invariant when we apply a certain kind of transformation, or as an equivalence class or type associated with some equivalence relation. The structures we recognize will be relative to our devices for specifying forms, or transformations or equivalence relations. Furthermore, by enriching or curtailing these devices we will obtain different notions of structure, count different things as having the same structure, and recognize different relationships between structures. Geometry illustrates this nicely. It recognizes broader and broader types of structure, starting with, say, congruent figures in an oriented plane, and moving on to similar figures in a Euclidean plane (without orientation), and thence to ones in an affine plane or a topological space.

Logic also deals with gradations of structure. Thus the sentences:

> If all animals are mammals, then all fish are whales;
> If something crashed then the baby cried,

have the same truth-functional structure but different quantificational structures.

This would be of passing interest to a theory of structures except that in Chapter 10 I used logical notions to define pattern occurrence, sub-pattern, and equivalence. Their definitions are relative to a background notion of definability. This, in turn, depends not only upon the non-logical vocabulary of one's background theory but also upon its logical devices. Indeed, not even the notion of isomorphism is absolute. For example, in a Euclidean space without a metric or orientation, similar triangles count as isomorphic, for there are invertible transformations of one triangle into the other which preserve all the properties defined on such a space. On the other hand, we can define Euclidean spaces with more structure in which non-isomorphic but similar triangles exist.

We should clarify the potential ambiguities in structuralist notions by distinguishing various occurrence, sub-pattern, and equivalence relations. Thus we might have:

> *P* first-order-occurs within *Q*,
> *P* second-order-occurs within *Q*,
> *P* set-theoretic-occurs within *Q*,

and other types of occurrence, depending upon the type of definability we recognize. We should similarly recognize different variants of the substructure relation and various forms of structural equivalence, since I also used the notion of definability in explaining these.

In a formal theory, isomorphism should be relativized to types of structure too. In ordinary mathematical practice, however, talk of isomorphism takes care of itself. For when we speak of two structures as being isomorphic we tacitly presuppose that the two structures are of the same type (for example, both are groups or both are metric spaces). Without this presupposition it would be unclear which features are to be preserved in qualifying a mapping as an isomorphism between the structures.

Not only do structural relations vary with our 'logical' background, but so do the things we take as exhibiting the same structure, and consequently the types of structure we recognize. If we limit ourselves to describing structures as the models of various first-order schemata, then the types of structures we will define will be like the more coarse-grained ones frequently found in abstract algebra. Here one starts by defining a type of structure such as a group, a ring, or a lattice with the intention of allowing for many non-isomorphic examples of the same type. As a result most of our structural descriptions will fail to be categorical. On the other hand, using second-order schemata, we can formulate categorical descriptions of the structures studied by (second-order) number theory, Euclidean geometry and analysis, and categorical extensions of ZFC that are considered powerful enough for most mathematical needs.

Thus, depending upon our logical resources, we might introduce:

The First-Order Natural Number Structure,
The Second-Order Natural Number Structure,
The First-Order Structure of the Reals,
The Second-Order Structure of the Reals,

and so on. By going to stronger logics we get more fine grained versions of the various structures. Thus the first-order examples listed above are not isomorphism types while their second-order cousins are.[8]

By fixing a background logic *L*, we can call an *L*-structure a structure in the unqualified or absolute sense. But this need not dispel all structural relativity, since by augmenting *L*'s logical devices we might make more intuitively structural distinctions than *L* does. Indeed, one objection to using first-order logic as the structuralist background is that both standard and non-standard models of number theory exhibit the First-Order Natural Number Structure.

[8] Those who hold that structures are isomorphism types could use another term to indicate that the first-order versions are structural properties instead of a structure. Likewise they might prefer other terms to 'The Group Structure', 'The Ring Structure', etc.

However, even second-order logic does not suffice for making all the intuitively structural distinctions that we might wish. For example, all and only progressions have the Second-Order Natural Number Structure, but if we fix a particular progression we can distinguish within it those sub-progressions in which the 'distance' between successive members remains constant from those in which the distance increases (for example, the even number sequence versus the prime number sequence). Thus while the structural notion of a progression is second-order definable, that of a progression with an increasing (or constant) distance between its terms is not.

One might propose eliminating structural relativity by taking *the* structure of the *X*s as the most finely grained characterization of the *X*s in terms of purely *logical* schemata. This, for example, would distinguish the absolute structure of the primes (that is, their being a progression) from their arithmetic 'structure'. Now I would find this suggestion persuasive if I felt that there were clear limits to the scope of logic and to the correlative notion of logical form. But I argued in Chapter 8 that the limits of logic are arbitrary, and that *logical truth*, *consistency*, and *implication* are not metaphysical properties and relations.

Another idea is for structuralists to take their background 'logic' to be the minimal one required for obtaining categorical axioms of number theory, geometry, and analysis. This idea fails, because the minimal logic necessary for formulating categorical axioms for these theories varies with the theory being considered. The logic of the quantifier 'there are finitely many', for example, suffices for number theory, but not for geometry or analysis. On the other hand, by using only free second-order variables we can formulate categorical versions of these three theories, but we still cannot make all the structural distinctions that we can make using bound second-order variables.[9] Thus the case for using one logic rather than another is not decisive, although second-order logic and set theory have an edge on the grounds of their expressive power.[10]

One might think that we structuralists are obliged to produce a precise theory of structures, if not a definition of *structure*, and that this will force us to settle on one type of structure as our official notion. But it is not clear that structuralism, or any philosophical

[9] See Shapiro (1991), ch. 9.

[10] Cf. the assessment in Shapiro (1991).

approach to mathematics, need be formulated as a theory, much less as a regimented mathematical theory. (I say more about this in Section 5 below.) Furthermore, there is no reason why a theory of structures could not recognize all L-structures for all choices of L.

Does this mean that a theory of structures should be based upon the weakest logic feasible, that is, first-order logic? Not necessarily. Adding non-logical mathematical devices to the first-order theory would allow it to make more finely grained structural distinctions. But a theory starting out with a stronger 'logic' could recognize the more coarsely grained structural distinctions of the first-order theory by introducing restricted notions of definability and structure. The moral is simply that since structural relativity is here to stay, a theory of structures should make a place for different types of structure.

5. STRUCTURALIST FORMULATIONS OF MATHEMATICAL THEORIES?

Using second-order logic one can formulate 'structuralist' versions of number theory, analysis, and Euclidean geometry. Given a sentence, $S(C_1, \ldots, C_n)$, in one of the usual second-order versions of these theories, we replace its non-logical constants, $C_1, \ldots, C_n$, by second-order variables to arrive at a second-order schema, $S(V_1, \ldots, V_n)$. Then we 'translate' the original sentence S as the universally quantified conditional:

$$(1)\ (\forall V_1), \ldots, (\forall V_n)(Ax(V_1, \ldots, V_n) \rightarrow S(V_1, \ldots, V_n)),$$

and take as our axiom the existential quantification:

$$(2)\ (\exists V_1), \ldots, (\exists V_n)Ax(V_1, \ldots, V_n).$$

(We obtain '$Ax(V_1, \ldots, V_n)$' from the conjunction of the axioms.)

The 'translations' (1), in effect, replace the statements S by statements asserting, roughly, that S holds in all structures of the appropriate type. Thus instead of saying that there are infinitely many prime numbers, the structuralist translation says that in any natural number sequence (omega sequence) there are infinitely many 'prime' positions. The structuralist axioms (2) can be construed as asserting that there are structures of the appropriate type. But note that they do not posit these structures as individuals.[11]

The second-order versions of number theory, analysis, and geometry have the virtue of being both finitely axiomatizable and categorical. Finite axiomatizability allows one to pack all the conditions describing a structure into the antecedents of the structuralist translations, while categoricity ensures their bivalence. For they are true in all the relevant structures or in none of them. Unfortunately, formulating structuralist versions of set theory is complicated by the fact that set theory is not categorical. Speaking imprecisely, we can finitely axiomatize the notion of *an* iterative hierarchy without thereby distinguishing between iterative hierarchies of different 'sizes'. This allows for the existence of many iterative hierarchies nested within each other. Thus to achieve categoricity and bivalence we must limit the sequence of nested hierarchies. The resulting categorical and bivalent theories are generally regarded as *restrictions* of set theory.[12]

So far we have been discussing mathematical formulations of structuralist axiomatizations of specific mathematical structures (or a class of structures in the case of set theory). These postulate as individuals only positions in the structures they characterize, and not the structures themselves. But we have not yet tried to develop a general theory of structures. By using a second-order logical background, we can go part-way without positing structures as individuals. For example, the notion of an omega sequence and an iterative hierarchy can be defined in second-order logic using two predicates, which I will abbreviate as 'S is an omega sequence' and 'E is an iterative hierarchy'. This allows us to form the sentence:

$(\forall S)(E$ is an iterative hierarchy $\rightarrow (\exists S)(S$ is an omega sequence & S is a sub-relation of $E))$,

[11] Hellman (1989) uses essentially this method in formulating modal-structuralist translations of standard mathematics. To anticipate applications of these theories, Hellman introduces a second-order 'domain' variable and restricts all first-order quantifiers to it. More significantly, he modalizes his translations by prefixing a necessity operator to sentences of the form (1) and a possibility operator to those of the form (2). The latter replaces claims that 'structures' exist by claims that they could exist.

[12] For more on these issues see Hellman (1989). For a discussion of the attitude of contemporary set theorists towards limiting the iterative hierarchy see Maddy (1988) and (1990).

which is a way of explicating the claim that every iterative hierarchy contains an omega sequence as a sub-pattern. We can also define isomorphism relations between specific types of structures, and prove that each is an equivalence relation.

But this approach seems to hit its limits fairly quickly. For example, although it can deal with the claim that every iterative hierarchy contains an omega sequence, it does not appear to have the resources for dealing with the claim that every iterative hierarchy contains *uncountably* many omega sequences. (Although we can define the *first-order* quantifier 'there are uncountably many' in second-order logic, this example requires an analogous second-order quantifier.) More significantly, I do not see how to develop the notions of pattern occurrence or pattern equivalence in purely second-order terms, because they use the metatheoretic notion of definability. Nor do I see how to express the idea that a pattern 'consists' of a domain of positions and relations on that domain.

Thus, it seems that developing a mathematical theory of patterns will require individual variables ranging over patterns and a sufficiently rich background theory for capturing the notions of pattern, pattern occurrence, and equivalence.[13] Taking set theory as a background theory might work. But since set theory cannot countenance the entire iterative hierarchy, this would omit an important mathematical pattern.

Unfortunately, the difficulty is quite general. For in treating patterns as individuals and describing their relations, we are treating patterns themselves as positions in a pattern of patterns. But this pattern, call it P, cannot itself be one of its positions. For suppose that P is a position of P, then it would be a position in an even bigger pattern, namely, the pattern whose positions are P itself together with the positions of P and whose relations are the relations of P plus the relation P's positions bear to P by virtue of being positions of P.[14]

[13] It may be enough for it to have the ability to give inductive definitions. Let the *definability closure* of a finite set of relations *R* be the set of relations definable in terms of members of *R*. Every relation definable from the relations of a pattern is a member of the definability closure of those relations, and, furthermore, it a finite logical construction of them. Thus it is likely that one can replace the metatheoretic notion of definability used in defining pattern occurrence and equivalence by an inductively defined notion of the definability closure of a set.

[14] I would not go so far as to call this argument a mathematical proof. For

In this regard pattern theory would be no worse off than set theory, which cannot recognize its own universe of discourse as a set. There is no getting around this—short of revising the fundamental principles of set theory. Yet by positing the existence of further extensions of the iterative hierarchy one can collect all the sets one previously recognized into a new set.[15] Pattern theorists might avail themselves of something similar. But they would have an options that most set theorists don't. For most set theorists are guided by the principle that every well-defined collection of objects is a set, which pushes them towards collecting the members of the iterative hierarchy into a set; whereas pattern theorists could claim that there is no fact of the matter as to whether the pattern of all patterns is a mathematical object.

We should also remember that a theory of patterns should suitably restrict excluded middle so that there is no automatic fact of the matter concerning the identities of positions in different patterns. This is required even on the second-order approach, since the sentence:

> Any position of an omega sequence is or is not identical to any position of an iterative hierarchy,

and others like it, are trivial consequences of excluded middle.

But I will end these speculations on the shape of a pattern theory here. I have neither the talent nor the taste for carrying out the project. Moreover, I see no current mathematical need for doing so. Mathematical model theory and category theory have already proved useful for studying structures in the guise of models of formal systems and categories. I would not expect a pattern theory to supply a framework superior to these.

6. THE STATUS OF STRUCTURALISM

If I am correct that there is no mathematical need for a theory of patterns, then the main motivation for developing one would be to

although its premisses are prima facie true, one might restrict them in developing a mathematical theory of patterns.

[15] It is not correct to say that the old set-theoretic universe becomes a set, for the universe does not change. Rather, after positing the extension one can prove that there is a set which contains those members of the iterative hierarchy whose existence we could previously establish.

show that my informal theory of patterns can be made mathematically precise. Now we have already seen some obstacles to carrying out this project completely. Contrary to one's expectations of a comprehensive theory of patterns, we have no assurance that we can consistently maintain that the pattern of all patterns is a pattern—in the sense of being one of its own positions. Formalizing my pronouncements on facts of the matter is likely to require the use of a non-classical logic. What is worse, some may resist formalization altogether. This raises the question of whether these limitations have a negative bearing on structuralism.

It would, if structuralism purported to be a mathematical theory rather than a philosophical account of mathematics. But, as I said in concluding Chapter 10, I did not put forth the slogan 'mathematical objects are positions in patterns' in order to herald a foundational programme, but rather in the hope of achieving a deeper understanding of the epistemology and ontology of mathematics. Despite this, I am uneasy about dismissing worries concerning formalizing structuralism simply on the grounds that it is not a mathematical theory. One reason is that I think that in philosophy we should strive for clarity, and the limitations on mathematizing structuralism might signal unacceptable limits to clarifying it. The other is that, as a holist, I recognize no place for a priori philosophy of science and mathematics. At the least, those parts of philosophy that purport to be descriptive should be continuous with science and mathematics. So I do not want to say that structuralism is philosophy, not mathematics, and leave it at that.

Let us consider the precision issue first. Formalizability within contemporary mathematics is arguably too strict a standard even for contemporary science, much less for structuralism. Many areas of science make essential use of counterfactual conditionals, causal statements, and vague predicates, none of which can be formalized within contemporary mathematics. I could cite other difficulties with formalizing science, but I don't think I need labour the point that if controversies about how to formalize a doctrine within a currently accepted mathematical framework were sufficient to condemn it, then little but mathematics would survive.

Furthermore, structuralism is not the only philosophy of mathematics that, on its own saying, eludes complete formalization. Consider realism about sets. The most widely accepted axiomatic set theory proves that the collection of all ordinals is not a set. Yet one of the guiding informal, philosophical principles of set theory is that any collection of objects that we can consistently describe forms a set. Obviously, this principle cannot be added as an axiom.[16] Or take the claim that all mathematical objects are sets. This can be proved trivially if construed as asserting that every set is a set. But how can one prove it non-trivially without begging the question? The same difficulty confronts the Church–Turing thesis that every effectively calculable function is recursive. Yet some texts take Church's thesis as a principle of informal, working mathematics.[17] The theses of structuralism are no worse off than the guiding principles of set theory, or the theses of set reductionism, or that of Church and Turing. So, once again, we would apply an unfairly high standard in dismissing structuralism just because some of its theses elude formal treatment.

Of course, a general discussion of this sort does not deal with the specific difficulties structuralism faces. Philosophers and logicians have convincingly defended the Church–Turing thesis against worries about its non-mathematical status. I should do the same for structuralism, and I tried to do so at the end of Section 2. In this section I have been trying to supplement my earlier discussion by considering the worry that while I may have shown how to make my ideas clearer, I still did not show how to make them *mathematically* precise.

I turn now to the consideration supposedly based on holism. To use this against structuralism one would need to argue that because philosophy of mathematics grades off into mathematics, a philosophy of mathematics should be capable of achieving mathematical precision. But when put in this way, I cannot think of why anyone would find it convincing. First of all, its conclusion does not follow from its premisses. If philosophy of mathematics grades off into mathematics, then the border between them will be blurred. If so, it may be unclear whether certain claims belong to mathematics or to philosophy of mathematics. (The Church–Turing thesis is a good example.) But it is entirely consistent with this that mathematical and philosophical claims distant from the border are recognizably different from each other, and that one way in which they differ is in precision.

[16] In fact, it is not clear how to formalize it!
[17] See Rogers (1968).

Secondly, the premiss is false. Much philosophy of mathematics—much of this book—is concerned with *general* ontological and epistemological questions that arise in connection with both mathematics and other disciplines. Its claims have little or no mathematical content, and it uses philosophical methods rather than mathematical ones.

This is often true even when mathematics is used in support of philosophy. Consider the philosophical claims about mathematics that have been buttressed by mathematical research programmes. Frege and Russell erected axiom systems to support their logicism, Hilbert developed proof theory to justify his finitism, and Brouwer initiated new mathematical methods as a result of his intuitionist philosophy. More recently, Hartry Field, Charles Chihara, and Geoffrey Hellman have done formal work in the service of nominalism and modal structuralism.[18] Here philosophy and mathematics do appear to blend. But a closer look at these examples reveals the formal work to be the product of philosophically motivated mathematical research programmes, whose success was necessary for the plausibility of *independently argued* philosophical theses. Frege, for instance, gave extensive philosophical arguments for the thesis that mathematics is logic. It is even likely that he understood that they were necessary complements to his mathematical work, since he knew how to reduce arithmetic to geometry. Field, to take another example, offered his constructions to show that mathematics is not indispensable to science, and based his nominalism almost entirely upon epistemological arguments.[19]

Finally, we should remember that holism is a thesis about scientific evidence. For holists there is no separating philosophy and science on evidential grounds. When a scientific theory conflicts with experience, holists hold, its philosophical and mathematical presuppositions are called into question along with the theory itself. To be sure, some statements may be as much philosophical as scientific or mathematical, but this point is an addition to holism, not a consequence of it.

In sum, I am for trying to make structuralism more precise as the

[18] And this list is by no means comprehensive.

[19] Ironically, philosophy and mathematics also appear to blend in this way in metaphysics, philosophy of language and epistemology, where formal techniques have been used in developing and studying possible world semantics and various 'logics', such as epistemic logic and deontic logic.

need arises. But like other philosophies it may be that the some of it must be regarded as a ladder ultimately to be kicked away. It is fruitful to compare mathematical theorizing to the study of patterns, but if all mathematically precise talk of patterns must be construed as talk of entities (and it is not clear that it is), then this part of the analogy must be given up.

7. STRUCTURALISM, REALISM, AND DISQUOTATIONALISM

Realism about mathematical objects does not commit one to realism about structures even when one maintains that mathematics studies patterns or structures and that mathematical objects are positions in patterns. This is simply because current mathematics does not affirm the existence of structures (as patterns rather than as positions). And I have been cautious about asserting the existence of patterns, even to the extent of holding that there is no fact of the matter as to whether the pattern number theory treats is a mathematical object.

Some philosophers who have found the idea that mathematics studies patterns attractive have wanted to take structural properties, construed as metaphysical universals, as primitive entities and interpret mathematics within a theory of universals.[20] Their mathematical realism is a consequence of their being realists about properties first. I am a realist about mathematical objects first without being a realist about properties at all, and in defending my realism I cite the philosophy of science (indispensability and holism) rather than metaphysics.

As a mathematical realist I am committed to the existence of mathematical objects, but I acquire that commitment through making the language of mathematics my own and acknowledging the truth of standard mathematical claims. Thus I am committed to the existence of numbers, functions, sets, and the like, and to mathematics' assertions about them. Furthermore, my holding that there is no fact of the matter whether numbers are sets, in no way impugns my mathematical realism. For in disallowing a truth-value to sentences such as 'Numbers are sets', I don't reject numbers or sets, etc., or the

[20] Several people have made this suggestion to me in conversation. It is the foundation for Bigelow (1988).

standard mathematical claims about them, or the thesis that these objects exist and these truths hold independently of us, our constructions, and our proofs.[21]

Despite this, my views on truth and facts threaten to collide with my structuralism when it comes to set theory. In affirming realism about sets I count various set-theoretic sentences as true. In particular, since set theory affirms excluded middle, I accept its instances in the language of set theory. But then my disquotational approach to truth commits me to the bivalence of set-theoretic sentences.[22] So far, this does not introduce any problems. However, we also know that set theory is not categorical: it does not pick out a unique isomorphism type. Axiomatic ZF set theory in its strongest (second-order) form characterizes a nested collection of iterative hierarchies rather than just one 'true' hierarchy of sets. By adding certain axioms, such as $V = L$, the axiom of constructibility, one can close off the nest and produce categorical and bivalent systems. However, most set theorists favour adding axioms that extend the known reaches of the iterative hierarchy instead of fixing its limits. None of these further extensions of set theory are categorical either.

Some philosophers and set theorists still think that set theory describes a definite iterative hierarchy, and that set-theoretic sentences are true or false according to whether they describe that hierarchy. But one would expect structuralists to hold that, on the contrary, (*a*) ZF set theory studies a family of hierarchies; (*b*) the sentences independent of ZF are *true in* some hierarchies and *false in* others; and furthermore, (*c*) there is no fact of the matter as to their unqualified truth. Unfortunately, part (*c*) of this position is incompatible with my apparent commitment to set theory's bivalence.

Let me also note that the view expressed by (*a*)–(*c*) also goes beyond mathematical realism—at least as I have been construing mathematical realism. For current mathematics has not taken a stand on the question of whether there is but one iterative hierarchy or many related ones.[23] The independence proofs produce models of

[21] The reader might want to compare this with Quine's affirmation of his realism on the heels of an argument for ontological relativity. See Quine (1981a), 20.

[22] In endorsing set theory I also affirm the appropriate instances of excluded middle in my metalanguage.

[23] This is not to say that *mathematicians* have not taken stands on both sides of the issue.

set theory in which this or that sentence is true and others where it is false, but they do not exclude the possibility that these models are all part of some unique iterative hierarchy. If so, the independent sentences would have unqualified truth-values in this larger hierarchy. Unfortunately, the local conception of mathematical evidence does not seem to dictate a choice between the one and the many hierarchies. Also, the matter is so far removed from contemporary natural science that taking a more global perspective is unlikely to prove decisive.[24] Thus my mathematical realism does not require me to hold that there are many iterative hierarchies. Nor does my structuralism, because it is a view about the nature of those objects which mathematics does recognize, and not a view about what it should recognize.

The bivalence question is complicated by first-order ZF set theory's having a much wider variety of models than second-order set theory. Second-order ZF set theory decides the continuum hypothesis in the sense that either it or its negation is a *semantic* consequence of the axioms of second-order ZF. (However, we are currently unable to establish which is the case.) On the other hand, the continuum hypothesis is independent of first-order set theory. Indeed, first-order set theory does not even characterize a nest of iterative hierarchies. Thus it is more plausible that second-order set theory is concerned with a unique structure than that first-order set theory is. Consequently, I will treat the bivalence of first-order set theory somewhat differently from the way I will treat that of second-order set theory.

However, before I turn to this let me note that although my realism does commit me to the existence of sets and the usual truths about them, it need not commit me to applying the truth-predicate to the entire language of set theory. Recall that for immanent realists the truth-predicate is largely a device for affirming each member of a class of sentences which is not finitely axiomatizable. Second-order ZFC is finitely axiomatizable. So is the multi-sorted system of von Neumann, Bernays, and Gödel, which is another extension of first-order ZFC. By affirming one of these theories I could endorse as much set-theoretic truth and existence as mathematics currently accepts while bypassing the issue of bivalence. I will not use this

[24] As I pointed out in Chapter 7, from a strictly holistic point of view, there is no need to make a decision.

dodge, but because the approach I will offer next is fairly tentative, I reserve the right to return to the dodge in responding to future difficulties.[25]

Let us turn to second-order ZF (and ZFC) first. Since its models form a nest, each occurs within all the hierarchies that it succeeds in the nest. Thus we can think of second-order ZF as studying a definite mathematical pattern: namely, the whole nested sequence of iterative hierarchies. (Of course, this nested sequence is not a set, and need not be a mathematical entity at all. See Section 2 above.) On this view, a set-theoretic sentence would be true or false according to whether it was true of this pattern, and bivalence would hold. Obviously, large technical difficulties would arise in working this out, since we would probably want to quantify over iterative hierarchies.[26] But my aim here is only to present a plausible way for structuralists to see second-order set theory as bivalent.

When it comes to first-order ZF, on the other hand, the question is whether to regard it like group theory or first-order number theory. Both theories have non-isomorphic models. We regard group theory as the general theory of all such models without singling out any as the standard group, and we do not count its sentences as true or false without qualification but merely true or false with respect to this or that interpretation. Some treatments take its formulae to be logical forms (schemata) of sentences, instead of sentences. Group theory does not posit a subject-matter; when it needs examples of groups it turns to other branches of mathematics. By contrast, we count first-order number theory as bivalent by taking it to describe an isomorphism type (or standard model). When pressed, we can specify this model further by using second-order number theory or set theory to define the natural numbers as the closure of zero under successor.

Now I don't think it makes much sense to construe first-order ZF along the lines of group theory. It is foreign to the practice of set theory, which instead of seeking its examples in other branches of mathematics posits its own subject-matter and uses it to construct

[25] Another way out is to hold that the current language of set theory is ambiguous.

[26] One idea is to count a set-theoretic sentence as true if it is true in a least one hierarchy and in all hierarchies containing ones in which it is true, and count it as false otherwise. Another is to use something like Hellman's Putnam-semantics. See Hellman (1989).

and re-construct examples for other branches of mathematics. Furthermore, if we took the group-theoretic approach to first-order ZF, it is likely that we would take the opposite approach to second-order ZF in order to secure examples for the former. To be sure, we can find parts of 'iterative hierarchies' in other branches of mathematics. The theory of finite sets can be modelled in number theory, for example. But the structural assumptions of even first-order ZF cannot be modelled in any of the standard branches of mathematics. Thus we would be forced to turn to second-order set theory or some other equally powerful foundational system, such as those given using category theory.[27]

But if first-order ZF is analogous to first-order number theory, we can secure its bivalence similarly. To do this, we regard first- and second-order ZF as describing the same pattern just as we regard first- and second-order number theory as describing the same pattern. Then we simply take a first-order ZF sentence to have the same truth-value it has in second-order ZF.[28]

8. EPISTEMIC VS. ONTIC STRUCTURALISM: STRUCTURALISM ALL THE WAY DOWN

In Chapter 6 I argued that there is no clear ontological distinction between mathematical and physical objects. Combining this with mathematical structuralism immediately suggests the idea that even tables and chairs, mountains and people, are positions in patterns. However, my view does not commit me to this view of ordinary objects. For it is consistent with my position that while subatomic particles occupy the vague region between mathematics and physics, tables and chairs are unquestionably not mathematical objects. In fact, this is something I want to maintain.

I did emphasize in Chapter 6 that under certain circumstances

[27] Number theory provides enough objects for arithmetic models of first-order ZF, but it does not have the strength to prove that ZF has a model.

[28] Although I have been speaking of full second-order ZF in the previous paragraphs, the position I have sketched here can be developed with respect to ZF with second-order free variables. Its models also form a nest of iterative hierarchies, but it is not committed to second-order entities. One could construe its free variables as schematic letters. However, instead of taking them in the usual way as standing for open sentences in the fixed ZF language, one would take them to also cover open sentences of further extensions of the language of ZF. See Shapiro (1991).

subatomic particles can be even more strongly incomplete than mathematical objects, since there may be no fact of the matter as to the identity of electrons in some region. Thus there might be circumstances in which there is no fact of the matter as to whether some electrons are constituents of one table, person, or mountain rather than another. But this alone doesn't imply that there is no fact of the matter as to whether the tables, persons or mountains are the same (or whether some table is identical to some mountain). For suppose that the tables are certain regions of space-time. Then there is a fact of the matter as to whether they are the same although there may be none as to whether an electron which was once part of one table is still part of it.

Thus my views on mathematics do not commit me to structuralism all the way down. On the other hand, structuralism all the way down aptly characterizes Quine's doctrine of ontological relativity. He comes close to expressing himself in this way:

> Reference and ontology recede thus to the status of mere auxiliaries. True sentences, observational and theoretical, are the alpha and omega of the scientific enterprise. *They are related by structure, and objects figure as mere nodes of the structure.* What particular objects there may be is indifferent to the truth of observation sentences, indifferent to the support they lend to the theoretical sentences, indifferent to the success of the theory in its predictions.[29]

There are, of course, significant parallels between mathematical structuralism and Quine's doctrines. Both attempt to account philosophically for the incompleteness of mathematical objects or, as Quine has put it, that they are 'known only by their laws'.[30] And both theories disavow facts of the matter concerning identities between objects of one branch of mathematics and those of another.

Despite these parallels, mathematical structuralism has a narrower scope than Quine's and a different motivation. Quine derives his structuralism by arguing that if we reinterpret science by permuting objects and making appropriate adjustments to the extension of its predicates we can preserve the truth-values of all its sentences

[29] Quine (1990), 31, my emphasis. See Quine (1981c), 20.
[30] Quine (1969a), 44.

including its observation sentences.[31] Thus the totality of science including the observation level is unable to distinguish between its isomorphic models. What is more, we cannot step outside science to distinguish between alternative models by appealing to direct referential connections between words and observable objects, because there are no such connections: 'it is occasion sentences, not terms, that are to be seen as conditioned to stimulations . . ..'[32]

Thus Quine ultimately bases his structuralism all the way down on his views on reference and meaning. By contrast, mathematical structuralism, including my own, finds its roots in philosophical remarks of Dedekind, Hilbert, Poincaré, and Russell, and Paul Benacerraf's provocative thoughts on the multiple reductions of arithmetic to set theory. It takes the thesis that mathematical objects are incomplete ('known only by their laws') as a datum and tries to explain it, and consequently it does not go as far as Quine's.

The first step towards using the mathematical structuralist perspective to arrive at structuralism all the way down would be to establish that ordinary objects are incomplete too. Now talking of the incompleteness of mathematical objects is a way of emphasizing that in their usual formulations our mathematical theories cannot even express answers to certain questions concerning the identity of the objects they discuss. We find similar gaps in our non-mathematical theories and in our overall theory of the world (assuming that we have one). Physics tells us many things about atoms, botany tells us many things about roses, but neither tells us exactly which atoms are to be counted as constituents of a given rose or whether the rose is a set of atoms or a mereological sum of them or something else. Biology tells us much about species; it does not tell us whether they are sets of organisms or mereological sums of them.

There are even gaps concerning who or what we are. And I am

[31] Quine is tacitly appealing to an elementary and well-known theorem about theories formulated in first-order logic. One might rightly object that many scientific theories have not been formulated in first-order logic. But the theorem in question can be generalized to any formal language with a standard compositional syntax and set-theoretic semantics, and thus to certain formalizations of modal operators and counterfactual conditionals. Of course, this still does not answer the objection, since it is far from clear that the totality of science can be formalized in these languages either. It simply shows that Quine's argument is quite general in its application. For further discussion see Resnik (1996).

[32] Quine (1990), 20.

not talking about our not knowing whether we are the ones who, in a drunken stupor, cursed the Dean, because there is a fact of such matters. I am talking about whether we are identical with the sum of our atoms or with the region of space-time we occupy or with our counterparts in other possible worlds. It seems quite reasonable to hold that in these cases there simply is no fact of the matter.

But if ordinary physical objects are incomplete, then it might be reasonable to explain this by applying structuralism to them, even to ourselves. On this view, just as 'number' refers to positions in one pattern and 'set' to those in another, our common-sense terms refer to positions in one pattern while terms for counterparts in possible worlds, sums of atoms, and alter egos refer to positions in other patterns. Just as the number pattern occurs in the set pattern, the common-sense one probably occurs within these patterns. And just as in the number–set case, a consequence of the pattern approach is the lack of facts of the matter about the identity of common-sense objects with these other objects.[33] Although one could motivate structuralism all the way down in this way, I am not prepared to advocate the view or this way of motivating it here.

Another important difference between Quine and most mathematical structuralists is that they do not seem to regard structuralism as an *epistemic* view. Yet Quine does, as the next passage shows:

> Now how is all this robust realism to be reconciled with the barren scene that I have just been depicting? The answer is naturalism: the recognition that it is within science itself, and not in some prior philosophy, that reality is to be identified and described.
>
> The semantical considerations that seemed to undermine all this were concerned not with assessing reality but with analysing method and evidence. They belong not to ontology but to the methodology of ontology, and thus to epistemology.[34]

Clearly eliminative and modal structuralists who aim to purge mathematical theories of their apparent commitment to mathematical objects are advocating ontological theses. But what of the structuralism propounded here? The thesis that mathematical objects are nothing but positions in structures certainly *sounds* ontological, and many people have read my previous writings in that way. One way of

[33] Stewart Shapiro airs similar ideas in the last chapter of Shapiro (1997).

[34] Quine (1981a), 21. See also Quine (1992).

developing an ontological structuralism is to take it as positing an ontology of structures, themselves seen as universals that can exist prior to their instances. Such a view might interpret each branch of mathematics as describing its own structural universal or universals.[35] In earlier sections of this chapter I have distanced myself from such a view.

Another ontological reading of my structuralism is to take it as positing an ontology of featureless objects, called 'positions', and construing structures as systems of relations or 'patterns' in which these positions figure. Under this construal, structures are like relations in extension whose *relata* are positions, whereas under the first construal they are like relations in intension. Where I have encouraged an ontological reading it has been in the direction of this second construal—especially when I speak of mathematical objects as being positions in patterns.

Now that Quine's remark has come to light it is important for me to make my position clearer than I have so far. Notice that I have not proposed structuralism as a foundation of mathematics nor as an ontological reduction, but have instead offered it as a philosophical view of the nature of mathematical objects. I have also conceded that the sentence 'mathematical objects are positions in patterns' is to be taken as a slogan which may elude mathematical formalization. Given the choice of reading structuralism as positing an ontology (of positions or patterns) or as concerned with the 'methodology of ontology' of mathematics, my previous remarks fit the latter, epistemic reading much better than the former ontic one. I prefer the epistemic reading, not only because it better reflects my position, but also because it is more economical. Another pleasant feature of epistemic structuralism is that if one extends it to people, one will not be committed to the view that we are literally positions in a pattern.

Of course, Quine is not saying that tables and chairs are literally positions in a patterns either.[36] When he talks of objects being nodes

[35] Cf. Bigelow (1988).

[36] Notice that on Quine's view the abstract–concrete distinction is immanent to an ontology. In our ordinary or face-value ontology, tables and chairs are concrete, because they are just tables and chairs. Yet Quine allows that this ontology can be reduced to one of just sets with there being no fact of the matter as to which is our real ontology. Cf. Quine (1981a), 17–18.

he is theorizing about reference in science; he is doing philosophy of language and philosophy of science. But when he takes science and common sense at face value he can certainly distinguish between concrete chairs and abstract positions in a pattern. The difference between the ontological versions of mathematical structuralism and ontological structuralism all the way down is that even from within the perspective of science and common sense it is plausible that mathematical objects are literally positions in structures.

I want to add a caveat to my epistemic turn. Characterizing my structuralism as epistemic may suggest to some that the incompleteness of mathematical objects is to be understood as the thesis that there really is a fact of the matter as to whether, say, numbers are sets, but it is an unknowable fact. This is not my view. The only unknowable facts that I have recognized are those we can credit to bivalence, and I have proposed restricting bivalence to avoid the facts associated with the incompleteness of mathematical objects. Epistemology enters the picture through motivating my disclaiming these facts: since nothing mathematics countenances would fix these facts if we countenanced them, I have opted for denying that there is anything to be decided, instead of enlarging the notion of evidence applicable to the putative facts.

9. A CONCLUDING SUMMARY

In this book I have presented a package of ideas whose major elements are (in order of appearance): realism, immanent truth (and reference), indispensability, epistemic holism, postulationalism, and structuralism. These doctrines are independent of each other. For example, one can be a realist about mathematical objects without being a structuralist, and conversely, a structuralist without being a realist about mathematical objects. Yet my positions on some issues raise questions concerning my views on others or rule out certain ways of working out the details of positions on other issues. I have addressed such concerns in recent sections, and throughout the book I have tried to combine my doctrines into a coherent system. I offer this concluding summary as another overview of this system.

Mathematical realism, on my view, is the doctrine that mathematical objects exist, that much contemporary mathematics is true, and that the existence and truth in question is independent of our constructions, beliefs, and proofs. I brought talk of truth into my formulation of realism simply as a device for asserting infinite or indefinitely specified bodies of mathematical sentences. As a result, I claimed, realists can be satisfied with an immanent, disquotational account of truth. But since mathematics countenances more objects than we can construct, and presumes more facts than can be established, realism's independence claim follows from the truth of mathematics and some obvious premisses about our abilities and the nature of constructions and proofs. Similar reasoning shows that most mathematical objects are abstract. Thus realists must confront the usual metaphysical and epistemological problems generated by recognizing abstract entities. In examining these, I concluded that nothing obligates the realist to provide a causal, informational, or interactive epistemology for mathematics.

My formulation of realism raises the question of how much mathematics the realist must endorse. To support the independence thesis and to preserve the usual distinction between realism on the one hand and finitism and constructivism on the other, I took realists to be committed to at least classical analysis. But the strength of the evidence supporting mathematics varies from branch to branch, and, consequently, so should the strength of the realists' commitment to the objects posited and the claims made within each branch. Thus they should be more strongly committed to complex numbers than to inaccessible cardinals.

The evidence for mathematical realism emanates from three sources. One of these is the local mathematical evidence for specific mathematical claims. Thus, when asked for evidence supporting the claim that there are infinitely many prime numbers, I would point first to proofs of that theorem. But even anti-realists can take proofs as evidence of something; I take them as evidence that the theorem is true. When asked why one should, I would take a more holistic perspective. I would begin by showing how the truth of number theory (and its axioms) is both presupposed and confirmed in many successful applications in science, practical life, and even within mathematics itself. Furthermore, I would cite the use of mathematics and mathematical methods in formulating scientific theories and developing their consequences as part of my justification for taking mathematical proof as evidential. My talk of successful applications might be construed as an attempt to transfer the evidence confirming individual scientific theories to the mathematics they contain.

Such evidence can be relevant, but I think stronger evidence comes from the necessity of presupposing the truth of mathematics in the enterprise of science. Thus my argument would take a pragmatic twist, and lead from the manifest rationality of practising science to the rationality of endorsing the local conception of mathematical evidence, and thence to the proof that there are infinitely many prime numbers.

This pragmatic holism is the second source of my evidence for mathematical realism. The third comes from comparing realism with other positions in the philosophy of mathematics. Realism is the default position in this field. For it takes the logical form and semantics of mathematical sentences more or less at face value, which in turn allows it to produce a relatively simple account of the apparently realist attitudes underlying mathematical practice and the use of mathematics in science. Thus it counts in favour of realism that it accounts for mathematics as well or better than other contenders in the philosophy of mathematics.

The pragmatic argument for realism requires anti-realists who accept its reasoning to show that we could do science without mathematics, or at least without presupposing its truth. But, as we have seen, the accounts of applied mathematics promised by several anti-realist programmes remain incomplete. Furthermore, anti-realist declarations of epistemological advances over realism are merely promissory notes, which anti-realists may not have the resources to fulfil.

On the other hand, showing that realism is more attractive and defensible than its anti-realist rivals would be of no avail if cogent objections to realism remain. Traditionally, the lack of a plausible epistemology has been enough for many philosophers to reject realism out of hand. I have met this deficiency by coupling my holistic view of mathematical justification and evidence to a postulational account of the genesis of our mathematical beliefs. As I emphasized, positing numbers and sets no more calls into question their independent existence than positing the planet Neptune or quarks does theirs. Immanent truth and reference complement my postulational approach by obviating the need for explaining how mathematical language 'hooks on to' an independent mathematical reality.

Structuralism both expands upon and fills in details of the previous accounts. Comparing mathematical objects to positions in patterns explains the incompleteness of mathematical objects, and thereby removes another objection to realism. The idea that mathematics studies patterns sheds further light on its methodology and epistemology, and explains why mathematical objects play no causal or information-transmitting role in the acquisition of mathematical knowledge.

BIBLIOGRAPHY

Albert, David Z. (1994), 'Bohm's Alternative to Quantum Mechanics', *Scientific American*, 270: 58–67.

Azzouni, Jody (1994), *Metaphysical Myths, Mathematical Practice* (Cambridge: Cambridge University Press).

Balaguer, Mark (1994), 'Against (Maddian) Naturalized Platonism', *Philosophia Mathematica*, 3rd Ser., 2: 97–108.

—— (1995), 'A Platonist Epistemology', *Synthese*, 103: 303–25.

—— (1996a), 'Towards a Nominalization of Quantum Mechanics', *Mind*, 105: 209–26.

—— (1996b), 'A Fictionalist Account of the Indispensable Applications of Mathematics', *Philosophical Studies*, 83: 291–314.

Benacerraf, Paul (1965), 'What Numbers Could Not Be', repr. in Benacerraf and Putnam (1983), 272–94.

—— (1973), 'Mathematical Truth', repr. in Benacerraf and Putnam (1983), 403–20.

—— and Putnam, Hilary (eds.) (1983), *Philosophy of Mathematics* (Cambridge: Cambridge University Press).

Bigelow, John (1988), *The Reality of Numbers: A Physicalist's Philosophy of Mathematics* (Oxford: Oxford University Press).

Blackburn, Simon (1984), *Spreading the Word* (Oxford: Oxford University Press).

Boolos, George (1984), 'To Be is to Be the Value of a Variable (or Some Values of Some Variables)', *Journal of Philosophy*, 81: 439–49.

Brouwer, L. E. J. (1913), 'Intuitionism and Formalism', repr. in Benacerraf and Putnam (1983), 77–89.

Burgess, John (1994), 'Non-Classical Logic and Ontological Non-Commitment: Avoiding Abstract Objects through Modal Operators', in Dag Prawitz, Brian Skyrms, and Dag Westerstähl (eds.), *Logic, Methodology, Philosophy of Science*, No. 9 (Amsterdam: Elsevier).

—— and Rosen, Gideon (1997), *The Subject with No Object* (Oxford: Oxford University Press).

Carnap, Rudolf (1956), 'Empiricism, Semantics and Ontology', repr. in Benacerraf and Putnam (1983), 241–57.

Cartwright, Nancy (1983), *How the Laws of Physics Lie* (Oxford: Oxford University Press).

Chihara, Charles (1973), *Ontology and the Vicious Circle Principle* (Ithaca, NY: Cornell University Press).

Chihara, Charles (1990), *Constructibility and Mathematical Existence* (Oxford: Oxford University Press).

Corcoran, John (1980), 'On Definitional Equivalence and Related Topics', *History and Philosophy of Logic*, 1: 231–4.

David, Marian (1994), *Correspondence and Disquotation* (New York: Oxford University Press).

Davies, Paul (ed.) (1989), *The New Physics* (Cambridge: Cambridge University Press).

Davis, Philip J. and Hersh, Reuben (1981), *The Mathematical Experience* (Boston, Mass.: Houghton Mifflin).

Dedekind, R. (1963), 'The Nature and Meaning of Numbers', in *Essays on the Theory of Numbers* (New York: Dover), 31–115, transl. of 'Was sind und was sollen die Zallen' (1888).

Detlefsen, Michael (1986), *Hilbert's Program* (Dordrecht: Reidel).

—— and Luker, M (1980), 'The Four-Color Theorem and Mathematical Proof', *Journal of Philosophy*, 77: 803–20.

Devitt, Michael (1984), *Realism and Truth* (Princeton, NJ: Princeton University Press).

Dood, J. E. (1984), *The Ideas of Particle Physics: An Introduction for Scientists* (Cambridge: Cambridge University Press).

Duhem, Pierre (1954), *The Aim and Structure of Physical Theory* (Princeton: Princeton University Press), transl. by Philip P. Wiener of *La Théorie physique, son objet, sa structure*, 2nd edn. (1914).

Dummett, Michael (1975), 'The Philosophical Basis of Intuitionistic Logic', repr. in Benacerraf and Putnam (1983), 97–129.

Etchemendy, John (1988), 'Tarski on Truth and Logical Consequence', *Journal of Symbolic Logic*, 53: 51–79.

—— (1990), *The Concept of Logical Consequence* (Cambridge, Mass.: Harvard University Press).

Field, Hartry (1972), 'Tarski's Theory of Truth', *Journal of Philosophy*, 69: 347–75.

—— (1980), *Science without Numbers* (Princeton, NJ: Princeton University Press).

—— (1982), 'Realism and Anti-Realism about Mathematics', repr. with additions in Field (1989), 53–78.

—— (1984), 'Is Mathematical Knowledge Just Logical Knowledge?', repr. with additions in Field (1989), 79–124.

—— (1986), 'The Deflationary Conception of Truth', in Graham McDonald and Crispin Wright (eds.), *Fact, Science and Value: Essays on A. J. Ayer's* Language, Truth and Logic (Oxford: Blackwell).

—— (1988) 'Realism, Mathematics and Modality', repr. in Field (1989), 227–81.

—— (1989) *Realism, Mathematics and Modality* (Oxford: Blackwell).

Frege, Gottlob (1891), 'Über das Tragheitsgesetze', repr. in Gottlob Frege, *Kleine Schriften*, ed. Ignacio Angelelli (Hildeshem: Georg Olms, 1967).

—— (1959), *The Foundations of Arithmetic* (Oxford: Blackwell), transl. by J. L. Austin of *Grundlagen der Arithmetik* (1884).

Giaquinto, Marcus (1992), 'Visualizing as a Means of Geometrical Discovery', *Mind and Language*, 7: 382–401.

—— (1994), 'Epistemology of Visual Thinking in Elementary Real Analysis', *British Journal for the Philosophy of Science*, 45: 789–813.

Gibbard, Allan (1990), *Wise Choices, Apt Feelings* (Cambridge, Mass.: Harvard University Press).

Glymour, Clark (1980), *Theory and Evidence* (Princeton, NJ: Princeton University Press).

Gödel, K. (1963), 'What is Cantor's Continuum Problem?', repr. in Benacerraf and Putnam (1983), 470–85.

Goodman, Nelson (1955), 'The New Riddle of Induction', in Nelson Goodman, *Fact, Fiction and Forecast* (Cambridge, Mass.: Harvard University Press).

Grosholz, Emily R. (1991), 'Problematic Objects between Mathematics and Mechanics', *PSA 1990*, 2: 385–95.

Hacking, Ian (1969), 'What is Logic?', repr. in Hughes (1993), 225–58.

—— (1982) 'Experimentation and Scientific Realism', repr. in Leplin (1984), 154–172.

Hale, S. C. (1988a), 'Against the Abstract–Concrete Distinction', unpublished Ph.D. dissertation, University of North Carolina, Chapel Hill.

—— (1988b), 'Spacetime and the Abstract/Concrete Distinction', *Philosophical Studies*, 53: 85–102.

Heath, Thomas (1981), *A History of Greek Mathematics* (New York: Dover).

Hellman, Geoffrey (1989), *Mathematics without Numbers* (Oxford: Oxford University Press).

—— (1996), 'Structuralism without Structures', *Philosophia Mathematica*, 3rd Ser., 4: 100–23.

Heyting, Arend (1956), 'Disputation', repr. in Benacerraf and Putnam (1983), 66–76.

Hilbert, D. (1902), *Foundations of Geometry* (Chicago, Ill.: Open Court), transl. by E. J. Townsend of *Grundlagen der Geometrie* (1899).

—— (1971), 'Letter to Gottlob Frege', transl. in E.-H. Kluge, *Gottlob Frege: On the Foundations of Geometry and Formal Theories of Arithmetic* (New Haven, Conn.: Yale University Press).

Hodes, Harold (1984), 'On Modal Logics which Enrich First Order S5', *Journal of Philosophical Logic*, 13: 123–49.

Horgan, John (1993), 'The Death of Proof', *Scientific American*, 269: 92–103.

Hughes, R. I. G. (1989), *The Structure and Interpretation of Quantum Mechanics* (Cambridge, Mass.: Harvard University Press).

Hughes, R. I. G. (ed.) (1993), *A Philosophical Companion to First-Order Logic* (Indianapolis, Ind.: Hackett).

Jubien, Michael (1977), 'Ontology and Mathematical Truth', *Noûs*, 11: 133–50.

Kitcher, Philip (1978), 'The Plight of the Platonist', *Noûs*, 12: 119–36.

—— (1983) *The Nature of Mathematical Knowledge* (New York: Oxford University Press).

Krakowski, I. (1980), 'The Four-Color Problem Reconsidered', *Philosophical Studies*, 38: 91–6

Krantz, David, Luce, R. D., Suppes, P., and Tversky, A. (1971), *Foundations of Measurement* (New York: Academic Press).

Kuhn, Thomas (1977), 'Objectivity, Value Judgment and Theory Choice', in Thomas Kuhn, *The Essential Tension* (Chicago: University of Chicago Press), 320–39.

Kyburg, Henry (1984), *Theory and Measurement* (Cambridge: Cambridge University Press).

—— (1990), *Science and Reason* (Oxford: Oxford University Press).

Lakatos, I. (1978), *Mathematics, Science and Epistemology*, ed. J. Worrall and G. Currie (Cambridge: Cambridge University Press).

Lavine, Shaughan (1992), review of Penelope Maddy, *Realism in Mathematics*, *Journal of Philosophy*, 89: 321–6.

—— (1994), *Understanding the Infinite* (Cambridge, Mass.: Harvard University Press).

Laymon, Ronald (1984), 'The Path from Data to Theory', in Leplin (1984), 108–23.

Leplin, Jarrett (ed.) (1984), *Scientific Realism* (Berkeley, Calif.: University of California Press).

Levi, Isaac (1980), *The Enterprise of Knowledge* (Cambridge, Mass.: MIT Press).

Levin, M. (1981), 'On Tymoczko's Argument for Mathematical Empiricism', *Philosophical Studies*, 39: 79–89.

Linsky, Bernard and Zalta, Edward N. (1995), 'Naturalized Platonism versus Platonized Naturalism', *Journal of Philosophy*, 92: 525–55.

Maddy, Penelope (1988), 'Believing the Axioms', *Journal of Symbolic Logic*, 53: 481–511, 736–64.

—— (1990), *Realism in Mathematics* (Oxford: Oxford University Press).

—— (1992), 'Indispensability and Practice', *Journal of Philosophy*, 89: 275–89.

Malament, David (1982), 'Hartry Field's *Science without Numbers*', *Journal of Philosophy*, 79: 523–34.

Niiniluoto, Ilkka (1987), *Truthlikeness* (Dordrecht: Reidel).

Parsons, Charles (1965), 'Frege's Theory of Number', in Max Black (ed.), *Philosophy in America* (Ithaca, NY: Cornell University Press).

—— (1979–80), 'Mathematical Intuition', *Proceedings of the Aristotelian Society*, NS, 80: 145–68.

—— (1986), 'Quine on the Philosophy of Mathematics,' in L. E. Hahn and P. A. Schilpp (eds.), *The Philosophy of W. V. Quine* (La Salle, Ill.: Open Court).

—— (1990), 'The Structuralist View of Mathematics', *Synthese*, 84: 303–46.

Poincaré, Henri (1913), *The Foundations of Science: Science and Hypothesis, The Value of Science, Science and Method*, transl. by George Bruce Halsted (New York: Science Press).

Putnam, Hilary (1967), 'Mathematics without Foundations', repr. in Benacerraf and Putnam (1983), 295–311.

—— (1971), *Philosophy of Logic* (New York: Harper).

—— (1973) 'Explanation and Reference', repr. in Baruch A. Brody and Richard E. Grandy (eds.), *Readings in the Philosophy of Science*, 2nd edn. (Englewood Cliffs, NJ: Prentice-Hall, 1989).

—— (1975), 'What is Mathematical Truth?', in Hilary Putnam, *Matter and Method* (Cambridge: Cambridge University Press)

—— (1984), 'Is the Causal Structure of the Physical Itself Something Physical?', *Midwest Studies in Philosophy*, 9: 3–16.

Quine, W. V. (1951), 'Two Dogmas of Empiricism', repr. in W. V. Quine, *From a Logical Point of View*, 2nd edn. (Cambridge, Mass.: Harvard University Press, 1980), 20–46.

—— (1962), 'Carnap and Logical Truth', repr. in Benacerraf and Putnam (1983), 355–76.

—— (1969a), 'Ontological Relativity', in Quine (1969d), 26–68.

—— (1969b), 'Epistemology Naturalized', in Quine (1969d), 69–90.

—— (1969c), 'Propositional Objects', in Quine (1969d), 139–60.

—— (1969d), *Ontological Relativity and Other Essays* (New York: Columbia University Press).

—— (1970), *Philosophy of Logic* (Englewood Cliffs, NJ: Prentice-Hall).

—— (1981a), 'Things and Their Place in Theories', in Quine (1981c), 1–23.

—— (1981b), 'What Price Bivalence?', in Quine (1981c), 31–7.

—— (1981c), *Theories and Things* (Cambridge, Mass.: Harvard University Press).

—— (1986), 'Reply to Charles Parsons', in L. E. Hahn and P. A. Schilpp (eds.), *The Philosophy of W. V. Quine* (La Salle, Ill.: Open Court), 396–403.

—— (1990), *Pursuit of Truth* (Cambridge, Mass.: Harvard University Press).

—— (1992), 'Structure and Nature', *Journal of Philosophy*, 59: 5–9.

Rawls John (1971), *A Theory of Justice* (Cambridge, Mass.: Harvard University Press).

Resnik, Michael D. (1975), 'Mathematical Knowledge and Pattern Cognition', *Canadian Journal of Philosophy*, 5: 25–39.

—— (1980), *Frege and the Philosophy of Mathematics* (Ithaca, NY: Cornell University Press).

—— (1981), 'Mathematics as a Science of Patterns: Ontology and Reference', *Noûs*, 15: 529–50.

—— (1982), 'Mathematics as a Science of Patterns: Epistemology', *Noûs*, 16: 95–105.

—— (1984), review of Crispin Wright, *Frege's Conception of Numbers as Objects*, *Journal of Philosophy*, 81: 778–83.

—— (1985a), 'How Nominalist is Hartry Field's Nominalism?', *Philosophical Studies*, 47: 163–81

—— (1985b), 'Ontology and Logic: Remarks on Hartry Field's Anti-Platonist Philosophy of Mathematics', *History and Philosophy of Logic*, 6: 191–209.

—— (1989), 'Computation and Mathematical Empiricism', *Philosophical Topics*, 17: 129–44.

—— (1990a), 'Beliefs About Mathematical Objects', in A. D. Irvine (ed.), *Physicalism in Mathematics* (Dordrecht: Kluwer Academic Publishers), 41–71.

—— (1990b), 'Immanent Truth', *Mind*, 99: 405–24.

—— (1992a), 'Applying Mathematics and the Indispensability Argument', in J. Echeverria, A. Ibarra, and T. Morman (eds.), *The Space of Mathematics* (Berlin: Walter de Gruyter), 115–31.

—— (1992b), 'A Structuralist's Involvement with Modality', *Mind*, 101: 107–22.

—— (1994), 'What is Structuralism?', in D. Prawitz and D. Westerstähl (eds.), *Logic and Philosophy of Science in Uppsala* (Dordrecht: Kluwer Academic Publishers), 355–64.

—— (1996), 'Quine, the Argument from Proxy Functions and Structuralism', *Philosophical Topics*.

Riskin, Adrian (1994), 'On the Most Open Question in the History of Mathematics: A Discussion of Maddy', *Philosophia Mathematica*, 3rd Ser., 2: 109–21.

Rogers, Hartley (1968), *Theory of Recursive Functions and Effective Computability* (New York: McGraw-Hill).

Sayre-McCord, Geoffrey (1988), 'The Many Moral Realisms', in Geoffrey Sayre-McCord (ed.), *Essays on Moral Realism* (Ithaca, NY: Cornell University Press), 1–23.

Shapiro, Stewart (1983a), 'Conservativeness and Incompleteness', *Journal of Philosophy*, 81: 521–31.

—— (1983b), 'Mathematics and Reality', *Philosophy of Science*, 50: 523–48.

—— (1989), 'Logic, Ontology and Mathematical Practice', *Synthese*, 79: 13–50.

—— (1991), *Foundations without Foundationalism: A Case for Second-Order Logic* (Oxford: Oxford University Press).

—— (1993), 'Modality and Ontology', *Mind*, 102: 455–81.

—— (1997), *Philosophy of Mathematics: Structure and Ontology* (Oxford: Oxford University Press).

Shelton, La Verne (1980), 'The Abstract and the Concrete: How Much Difference Does This Distinction Make?', unpublished paper delivered at the American Philosophical Association, Eastern Division Meetings.

Shimony, Abner (1989), 'Conceptual Foundations of Quantum Mechanics', in Davies (1989), 373–95.

Sober, Elliott (1982), 'Realism and Independence', *Noûs*, 16: 369–85.

—— (1993), 'Mathematics and Indispensability', *Philosophical Review*, 102: 35–57.

Steiner, Mark (1975), *Mathematical Knowledge* (Ithaca, NY: Cornell University Press).

Teller, Paul (1980), 'Computer Proof', *Journal of Philosophy*, 77: 797–803.

—— (1990), 'Prolegomenon to a Proper Interpretation of Quantum Field Theory', *Philosophy of Science*, 57: 594–618.

Tharp, L. (1975), 'Which Logic is the Right Logic?', *Synthese*, 31: 1–31.

Tieszen, Richard (1994), review of Penelope Maddy's *Realism in Mathematics*, *Philosophia Mathematica*, 3rd Ser., 2: 69–81.

Tymoczko, T. (1979), 'The Four-Color Problem and its Philosophical Significance', *Journal of Philosophy*, 76: 57–83,

van Fraassen, Bas (1980), *The Scientific Image* (Oxford: Oxford University Press).

—— (1985), 'Empiricism in the Philosophy of Science', in Paul M. Churchland and Clifford A. Hooker (eds.), *Images of Science* (Chicago, Ill.: University of Chicago Press), 245–308.

Weston, Thomas (1992), 'Approximate Truth and Scientific Realism', *Philosophy of Science*, 59: 53–74

Wilder, Raymond (1968), *The Evolution of Mathematical Concepts* (New York: Wiley).

Will, C. (1989), 'The Renaissance of General Relativity', in Davies (1989), 7–33.

Wright, Crispin (1983), *Frege's Conception of Numbers as Objects* (Aberdeen: Aberdeen University Press).

INDEX